LET ME NOT BE MAD

A. K. BENJAMIN

LET ME NOT BE MAD

THE BODLEY HEAD
LONDON

1 3 5 7 9 10 8 6 4 2

The Bodley Head, an imprint of Vintage,
20 Vauxhall Bridge Road,
London SW1V 2SA

The Bodley Head is part of the Penguin Random House group
of companies whose addresses can be found at
global.penguinrandomhouse.com.

Penguin
Random House
UK

First published in the UK by The Bodley Head in 2019

www.vintage-books.co.uk

A CIP catalogue record for this book is available from the British Library

Hardback ISBN 9781847925428
Trade paperback ISBN 9781847925435

Printed and bound in Great Britain by Clays Ltd, Elcograf S.p.A.

Penguin Random House is committed to a sustainable future for
our business, our readers and our planet. This book is made
from Forest Stewardship Council® certified paper.

For my daughters

Author's Note

In order to ensure that no real person is identifiable from these pages and that any sensitive material on which I have drawn is protected, I have changed names, physical features, backgrounds, locations, genders, nationalities, ethnicities and the key detail of events. None of the people described are individuals but each is a blend of different encounters, real and imagined. If anything, this confusion makes the book more faithful as an account of my experiences.

A therapist who reaches the maximum level of empathy would become the patient: the two points of view would become fused.

Valeria Ugazio

Who gives anything to Poor Tom? Whom the foul fiend hath led through fire and through flame, through ford and whirlipool, o'er bog and quagmire; that hath laid knives under his pillow, and halters in his pew; set ratsbane by his porridge ...

William Shakespeare, *King Lear*

I have hundreds of such memories and from time to time one of them detaches itself from the mass and starts tormenting me. I feel that if I write it down I'll get rid of it.

Fyodor Dostoyevsky

You

We've become used to the cameras, TV crews of young men and women with hangovers, haircuts like Frank Gehry buildings, smoking in ambulance bays, idling outside theatre doors, waiting for the next emergency. At first our shirts got crisper, sleeves rolled above the elbow in keeping with Trust policy, and our intonations grew gentler, our questions tender, our eyes sought our patients' for the first time in years. 'Reality' was contagious. Nobody was immune, most were naturals. Then it drifted.

Now we file in stage left, the young, the old, the earnest, the furious, the hopeful, the guilty; men, on the whole, though many of us look and sound like boys, most in suits – the occasional throwback pince-nez or bow tie – a few in scrubs and running shoes; here we come, bound together by some singular idiomatic force, like a caste. We no longer notice the cameras, but it's still an act, our entrance a piece of theatre, and there would be no theatre without you, our audience: we file in to collect *you*.

Rifling medical notes we have only just opened, we bark out names as questions.

'Miss Jennifer Almendy?'

'Mr Konrad Kuchzynski?'

'Dr Mohammed Mosham Alawi?'

We often get them wrong, especially these days, in London. Hopefully one of you will hold up a hand and pull yourself to your feet with the help of a partner or a stick, or wheel yourself forward. But it's not uncommon for nothing to happen, for names to die unclaimed, our letters misplaced or not sent or never written, appointments dodged or forgotten by you – which might be symptomatic here in General Neurology. Worse still, you're here but it's too late: you've lost your speech, can't raise an arm, no longer recognise your own name.

There you are walking across the waiting area to meet me, gait normal. I introduce myself with an ease that's practised – smiling, holding out a hand, 'Please, just call me Ally' – hoping to relax you (sometimes it has the opposite effect). I lead you down a long corridor, lined with doors that say 'Epilepsy', 'Neuro-oncology', 'Multiple Sclerosis', 'Chronic Pain', 'Neuro-degenerative Illness' . . . Most of you fall silent. I may ask about your journey, if you found us alright, whether you've had a cup of tea. You might answer with thoughtless politeness, or start a story you can't stop, or you don't hear, your mind somewhere beyond the infernal corridor's end.

My room is the last on the left. I say mine but I have no home: different days, different rooms. Inside there are no photographs of children or dogs, no Renaissance print of captivated medical students gathered around an exposed cerebrum, no studiedly soothing abstract. Instead, bone-white walls, cold blue trim, strip-lighting, a dull grey desk, a rudimentary chair on each side, a second smaller desk in the corner on which stands a hefty, ageing computer and its monitor (a box in the corner of the screen shows E. coli incidence rates clicking upwards in real time), filing cabinets, shelves with a few outdated textbooks or journals that haven't yet been 'borrowed' forever. It might be an interview room in a police station. Years ago we asked Security to install a panic button if not CCTV in case of attack. Nothing happened. Then a red button appeared overnight which makes no sound anywhere when depressed.

We sit facing one another across the desk, just looking, the space between us charging. You are younger than most. Your silver hair is still wet. There's a whiff of cigarettes mixed with chlorine. The fading impression of goggles like quotation marks around your bright, grey-blue eyes. Eyes: water-bubble light-lassos, the heart's stigmata. At some unspoken level what needs to happen takes a moment, that's all it requires to know the channel is open, that I can *see* you and everything you bring, however devastating.

The moment passes, then another, and another ... There was a time when I wouldn't have thought of this as love, but what else could it be?

Before we met there was a more formal introduction; the referral letter. Usually brief notes dictated between professionals with you copied in. At once explicit – cold-heartedly so – and utterly bereft of detail. Not to complain, I know how busy we are. Yours mentioned 'general memory difficulties', a few 'unusual behaviours' without elaborating 'unusual', hinted at 'a lot of things going on' in your life, as though lives weren't meant for that. Now in the flesh, tumbling wet hair darkening the shoulders of your silk shirt, the letter laid openly out on the desk, I wince: the casual off-the-rack descriptors – your 'attractiveness', your 'resilience', your 'charm' – as though the last hundred years hadn't happened ...

Maybe you don't register this, your mind searching for the letter's terminus, which comes suddenly, out of nowhere, mid-sentence: 'whether all this constitutes a psychiatric overlay, likely stress-related? Or is it indicative of incipient organic process?'

In other words, anxiety or terminal disease?

Our questions to each other often hinge on such choices, known as 'differentials', usually pairs of names opposed like bouts: Alzheimer's vs Vascular? Alzheimer's vs Depression? Alzheimer's vs Lewy Bodies? Lewy Bodies vs Parkinson's? Parkinson's vs Progressive Supranuclear Palsy? Progressive Supranuclear Palsy vs Corticobasal Degeneration? And so on. Big names. Title fights. And, on the undercard, the exotica: Prion's, Guam, Hiroyama's Syndrome, Erb's Palsy, Gaucher's disease, Sanfilippo's, Rasmussen's, Dandy-Walker's, Wallenberg's, and hundreds, thousands more, an international telephone directory of doom. At its most elemental that is all we are; glorified receptionists, singling out one name from innumerable others, your name, that is, your *new* name, occasionally with an old-school personal touch, but more often these days as though from a call centre in Hyderabad.

Do you see through the argot? You do; it's in your eyes. It's in your eyes because it's in mine. What we want to know is if you have the beginnings of a terrible illness, at your young age most likely something that would devastate your functioning within three to five years, destroy you within seven to ten ... Or are you just being a human being?

It's up to me to actually *talk* to you. I am interested in everything: how you speak (with a light, occasional lisp), your syntax, lexicon, prosody, the quality of any jokes (you're a natural), your concentration, mental flexibility, temperament, self-awareness, the way you do or don't look me in the eye (you do, unfalteringly). I will explore your mood, your day-to-day functioning, your work, your leisure ('swim like a fish'), the books you read (Grace Paley, Pema Chödrön, never-ending Proust; 'a life sentence'), how you read them (standing up, eating crisps), the music you listen to (brass bands, Hank Williams, Radiohead) – sometimes I get carried away with questions, stray beyond obvious clinical relevance, but then curiosity might save the cat – your faith, your doubt, your dietary preferences ('eat like a fish' too?), your eyesight, hearing, sense of taste, smell, touch; together, we'll map your medical history, your family's (an aunt had semantic dementia), your relationship history, your sexual appetite ('intermittent fasting'). Then, most crucially, what, if anything, has *changed*. And if so, when? And in what way? And are you sure? And how you noticed? And how can you be sure? And has anyone else noticed? Really? Are you quite sure?

I hold it all, as best I can, along with everything else that stirs in me as you spill out. Your face: the eyes, almond-shaped like a Russian icon, the line of your nose, that lickerish mouth; one moment tight with fear, overtaken the next by fountainous laughter … Always faces, tens of thousands across a career, each one made up of countless micro-expressions which register everything; more liminal than a blood test, less decisive than a lumbar puncture, but meaningful all the same.

I can look to the medical Literature for help:

The multifaceted emotional dimensions of patient messages are often occluded in consultations (Lever & Segnit 2016), especially by male physicians who tend to rely on primitive rational decision-making tools even when they may be misleading (Hammond & Francis 2018).

Highburger & Kroll (2017) recommend careful attention be given to facial expression: eye contact, posture, head-nodding, hand gestures, sublingual 'umming' and 'uhuhing' …

Morrow et al. (2019) reckon at least 80% of communication is in observable non-verbal behaviour, particularly facio-manual gesturing.

Often the research points to only the coarsest shallows of what we do; our bodies and minds endlessly encoding one another in the most delicate, ineffable detail. Sometimes clinical need makes us stray into poorly lit places: like measuring 'intuition' – an 'old whore making yet another comeback', according to one recent paper; but however we define it, intuition is no more than a password for the small fraction of data we can't quite keep up with, beyond the even smaller fraction that we can, as the sum of one another, of you and I, works through us with limitless nuance. Of course there will always be the question of where to draw the line, how to guard against our own overly idiosyncratic or recondite interpretations: like our patients', our minds go everywhere.

The only difference is that you are compelled to disclose where yours is going.

'So, apparently I'm "charming" … and "resilient"; thank God.'

'Sorry about that.'

I jot down a note: hiding behind writing. A drop plashes onto the referral letter; from your hair. Your eyes? Maybe you use the same pool as me, though I haven't seen you before, and I've been there constantly, pounding out the lengths as my Swim Programme builds.

'At least now I've got one thing in common with Iris Murdoch,' you say. 'Could it have anything to do with going grey early?'

Dread has many fronts, irony included. You tell me you forget where you park the car at Waitrose, often leave your key in the front door, got lost once picking up the kids from school. Then you jag the other way, minimising what you've just told me, say you've always been 'ditzy', are famous for it. But your face mutinies, a fasciculation (tremor) in one corner of your mouth – the bluffer's tell.

'Besides, they've all got it in for me at home.'

There are so many different ways to hide; the 'fatalism' that I'm noting is likely a subtle way of pre-emptively legiti-mating a poorer than expected performance. But just as I catch up with you, you jump again:

'What do *you* think is happening to me?' like a child in its directness.

I put my pen down; looking means listening.

Despite your best efforts to keep things un-patterned and therefore dismissible, you can't do it: you think your brain is rotting. My intuition is you're right, but I can't tell you that, not yet.

What gets measured gets done, as with any business. I subject you to three hours of neuro-psychometrics ('Three hours! My second labour didn't last *that* long'). Measuring the mind is contentious; philosophically, ethically, statistically. It has faux-scientific roots in nineteenth-century Caucasian supremacy: comparing the volume of the skulls of different races by filling them with lead gunshot is one example. Over the years gunshot was replaced by puzzles, pictures, word formations. The profession always looks rather conservative, neurotic, whimsical and certainly anachronistic next to its glittering counterparts in nuclear imaging. Apologists speak of it as an art as much as a science, which is an admission of defeat where I work.

Now, somewhere in the recesses of hospitals across the country men and women (white Caucasian usually) in thick glasses fret over the validity of their instruments; how their functionality relates to different cognitive domains; how the domains themselves relate to different neuroanatomical loca-tions, so-called 'Brodmann areas', and so on ... It's been the decade of the brain for the last twenty-five years. Fashions come and go; the cortex, the subcortex, white-matter tracts, relay speeds, gamma oscillations, secret pathways which open only in the dead of night. Hegel wrote of a king obsessed with making a perfect map of his realm, perfect in its own right that is, regardless of the territory. Completed by map-makers to his satisfaction, he gave orders for armies of labourers and craftsmen to refashion the entirety of the kingdom – stone for stone, tree for tree, blade of grass for blade of grass – in the image of the map, until they ran out of money. It's not too cynical to suggest that we too might run the risk of getting lost constructing fantastically elaborate and expensive simulacra of our own ideas.

Especially when the ideas are as slippery as 'you'. Am I talking to a *charming* homunculus sitting somewhere behind an inordinately complex control panel; a pair of blue-grey eyes behind the painting's eyes? And would that 'you' have a smaller, more defenceless, even more charming you residing within, like a Russian doll? Or are 'you' fundamentally not there, there as only effect, an attractive epiphenomenon of the author-less integration of different faculties? An accident, in other words. It's the so-called Binding Problem. We never can quite make up our minds philosophically.

But the work goes on. I begin with an estimation of your 'premorbid' or baseline ability. If someone has a devastating brain injury and they are sitting in front of you, then however obvious the apparent effects, some anchor in what they were like before is required. Common sense may help; the level of educational and professional attainment, the reference of friends and relatives, etc. But common sense also suggests that these may be as misleading as helpful; how many mothers and fathers would *accurately* describe their children's optimal ability? So we use tasks based on 'over-learned' skills, such as vocabulary acquisition, which are more resistant to the effects of injury and disease than other abilities.

'I don't understand,' you interrupt.

'Bear with me.'

With such a mark in place, I may evaluate your current performance. Each neurological disease affects different neuro-anatomical locations and these areas have come, via the shrapnel in a soldier's brain or the surgical ablation of a monkey's, to be associated with discrete cognitive faculties. So I will look for those aspects of your performance which seem to be significantly different from the premorbid estimate.

'Does that make sense?'

'Premorbid? Meaning I'm not dead yet?'

'No.'

'I don't get it.'

It's OK, you don't need to, it's not your job. My job, including this obfuscatory speechifying (we call it 'psycho-educating'), is to use whatever I might call clinical expertise to decide whether your profile is *normal* or redolent of a pathology.

'OK?'

'OK.'

Part of you is defeated before we've begun. I press on, the clinic room booked by another doctor from midday.

'I'm going to give you a list of words and when I've finished I want you to tell me as many of the words as you can. OK?'

'In the same order?'

'Any order you like.'

'Right-o.'

'If you get stuck, try not to panic. We'll go over it again. If you need a break, or you want to ask me anything about the tests, or anything else – anything – just stop me. OK?'

You nod.

'Otherwise I'll plough on and on, like a robotic maths teacher. Ready?'

'Ready.' In a computer voice.

'Nail clipper ... Chile ... pomegranate ... moose ...'

I read them to you deliberately, without inflection, at a rate of one every two seconds, sixteen in total; as limpidly abstract as a John Ashbery poem. It's a supra span-task, too much for your hippocampal formation (the anatomical area associated with memory) to encode in a single trial. If your brain is OK it will find a way to learn:

'... lime ... tweezers ... scalpel ... Denmark. Over to you.'

'Right ... it was a cold day, in ... *Belgium*? She ate the grapefruit, or she was forced to eat it segment by segment with a pair of tiny tweezers, by her father who was holding a scalpel.'

I think how we can't resist this narrative urge, assembling bits and pieces as though they were fragments of something larger, like turning 'living' into 'a life'.

'I don't know ... Did it start raining?'

'I'm afraid I can't say.'

'Of course you can't.'

I want to help you.

'Here it is again. Remember; it's just a list of items, nothing more.'

Reciting the list a second time I notice how the curls have magically sprung from your drying hair, released like a genie from its bottle, getting wilder as the moments pass. My heart rate has increased by five to ten beats per minute, my breath – the proximal vane for all we think and feel – now shallow and quick, willing you to do better.

Your performance might just be 'ditziness', a temperamental artefact. Or it might just be a function of your peregrine history. You smoked a lot of grass and 'a little' heroin during the years you lived on the road, in Kerala, in northern Pakistan, on the west coast of Guatemala. You drink far too much white wine now. There are a number of current stressors: a daughter with bulimia; a live-in father who needs caring for after a stroke; a remote husband ('No, he's not aggressive, that would take too much energy'); mounting insomnia. These could only have a negative impact on your performance, so how reliable is what I'm seeing? And that's not even considering whether what you're saying is true, or completely true: studies have shown that your generation, our generation, lie on average two or three times every ten minutes, men to make themselves look better, women to feel good. So how many times would that be in an average assessment? Fifty? A hundred? Maybe more, when feeling good is a matter of life or death. So which bits are the lies? What do you most want me to see, most want to conceal? The same applies to me: I want you to see me in a particular way – warm, understanding, expert, handsome (in my own way) – and conceal from you everything that doesn't fit. I want to reassure you, tell you exactly what you want to hear, lie in the name of health. I think of Kafka's country doctor, dismissively convinced there is nothing wrong with his languid-looking patient, telling his family as much, only to find on closer inspection a terminal, worm-infested wound on the young man's flank.

The second and third attempts are no better than the first. On the fourth trial you recall two new items and lose one you'd remembered previously.

'I've done terribly, haven't I?'

'I'll score things up later.'

'Bottom of the class?'

'It takes a while. Let's move on.'

Your eyes glisten. I can see the pulse in your neck.

'I've always hated tests. My friend told me to arrange things like they're in a story.'

I gently nudge the box of Trust tissues towards you. I want to do more.

'Silly cow ... I'm still good at drawing.'

Thirty minutes later I try you on the same list, holding my breath this time, as I did as a child, underwater in the bath, between lamp posts on car rides, during a reprimand, or the sound of someone eating an apple: a practice in how to do without, and in willing the world to do my bidding ...

But no: awful again.

We are skull-jumpers, there is no limit to our identificatory capacity. Your face, voice, breath continue their unfolding, each now different from the last, changed beyond recognition in the two hours since we first met. Looking: more intimate than any physical examination.

The voltage switches once more, symptoms pooling between us, tributaries of some larger untold story – a comedy, a murder, a romance. Butterflies, heart rate trilling with frightening speed: 145 beats per minute, 150, 155, edging zone 4; breath galloping, hands trembling, thoughtlessly mirroring your posture, sculpting sympathy from the body's clay ... The intensity of neuronal activation is processed through deep limbic structures, and when intensity exceeds a genetically defined threshold it leads to activation of the autonomic nervous system, triggering unconscious automatic changes in cardio-vascular and respiratory systems – readying us to fight, to flee, to freeze, to love. We are sensitive, we have no choice.

Look at you: you are no accident. For once my dinosaur colleague is right: you really are 'charming'. We have spent the whole morning together. That's more time, with more tender, dedicated attention, than either of us would share with our children or partners in any given week.

Could willing them to do well (because for some more or less intractable reason the patient moves you with their meekness or thrills you with their presence) or willing them to do badly (because they take too much time, are foreign, ungrateful, insistent or because you have a headache, a parking ticket, a darkening lump under your armpit) affect things, things which aggregate and cohere in a way that might turn 'nothing' into 'something'? Or vice versa? We may never know. For now at least the conversation isn't possible because there is no medical language that could admit how feeling might possibly get in the way of facts rather than blindly serve them. There is intuition, but what I am talking about is way below intuition. From down there, in the darkest reaches of the ice, feelings are all we have.

'Peon ... palaver ... piss ... piss-flaps?'

We have gone beyond John Ashbery. Outside a pigeon is trying to land on the window's narrow, anti-pigeon ledge.

'Pontefract ... prick ... pathetic ...'

'That's time, I'm afraid.'

As many words as you can generate in a minute beginning with the same letter, excluding proper nouns and swear words.

'Oh God. Is this what "incipient organic process" looks like?' you ask. 'Am I post-morbid?'

In the time it takes to deflect your question I see your future unspool. Forgetfulness first: losing your children's friends' names, what you'd come into the living room for, what time you put the roast on, asking the cleaner if she's fed the cat, asking again five minutes later. The beginnings of 'dyspraxia': a moment where you've forgotten how the remote for the television works, which way the key turns in its lock, how the buttons on your blouse fasten. ('Ditzy?') The onset of 'anomia' following the rule of frequency: losing the name for Caerphilly, then Cheddar, then cheese, then children, your children. A steady upsurge of confusion: why the weekend started on Tuesday. Where the living room is. ('Don't we *live* in all our rooms?') If the cleaner has roasted the cat. What a key is for. (Still 'ditzy' by the skin of its teeth?) A fear of spending too long out of the house, of being in company, of being left alone. An obsession with fizzy sweets, with sauerkraut and satsumas, then the sudden death of appetite. Occasions when you can't make it to the toilet in time, or make it to the toilet to find it's moved, again. Pretending to read, stopping pretending. Paranoid that the increasingly simple crosswords your husband left out are deliberately impossible, or coded by him with retaliatory messages. Unable to write your own address, or read the reminders you've written for yourself, or know what a pen is for. Until sitting in the living room hour after hour, without the TV, the radio, a mobile phone – 'ditzy' long dead – and still almost totally distracted from the disappearing sense that you are losing everything including your self and those around you who, despite themselves, gripe and nag and grouse. *Almost* ... You are looking at me without blinking, eyes burning. I want to give you something back, but I can't.

When a clinician turns his hand to writing case studies, he might as well be writing fiction for all the relevance it bears to what actually happens in the room. If, like Philip Roth says, you're wrong about someone before you meet them, wrong again when you meet them, wrong once more when you think about the meeting afterwards, then when it comes to writing them up you might as well be working from scratch. Names and details must be changed, but we shouldn't stop there. Some fictionalising is inevitable; necessary and contaminating in equal measure: imagination might just help the next patient.

But this, here, really is *you*, unimagined, unwritten as yet. We are facing each other across the large grey desk; the unfolding won't stop: you, *the thing itself*, unadorned, powerless, supplicant, quietly devastated and somehow radiant. Unseen, a chain of rogue proteins grows to shadow what is brilliant white on contrast imaging.

'I need to see you again.'

'When?'

'In twelve months.'

'OK.'

'Maybe nine.'

'OK.'

'Reception will make an appointment.'

I'm dragging things out, hoping you will hear what I cannot say.

'Good.'

'Good.'

'I'm supposed to do sudoku, aren't I?'

'Only if you like sudoku.'

'No.'
'So no.'
'OK. Good.'
'Good.'
'Thank you. Good. I'm sorry.'
I'm sorry.

JB

'He has a high pain threshold,' says Milner, the father, smiling (for some reason) through a beard thick as swarming bees.

'No he hasn't. He can't stand pain.' The mother is sullenly, girlishly perched on the sofa's corner.

'He's a baby,' she tells me.

'He's nearly ten.'

'A big delinquent baby.'

'He's precociously adolescent.'

'Bit like his father.' She will only look at me.

'But more like his mother.' He will only look at me.

I am supposed to be in the middle. Really I am not there. I pretend to note down something significant:

The two next to one another, his-and-hers sheepskins, the snow on their boots not yet melted, a world between them.

'May I ask why he isn't here?'

'He refused to come, Doctor.'

'You let him refuse,' she says.

'Words, words, words.'

'That's right, w*ords*; what else would they be?'

DO A
NOT THA
00129/ 009 J/4/4
14:05 17/0

Páiste
Ch ld
€1.00
C/S

B/R 67

Merr on
C/

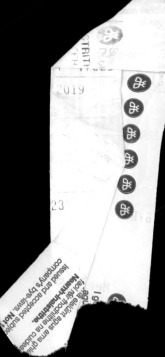

Less than two minutes into an hour-long interview, less than a month into a new career: it isn't just the patients who are vulnerable. I hadn't slept much in two nights wondering what he would be like in person. He'd been referred for a preliminary assessment to the Child and Adolescent Mental Health Services (CAMHS) where I was newly based. My mouth dried reading the GP's thoughts on possible diagnoses – ADHD, autism, conduct disorder, juvenile epilepsy, undiagnosed trauma, more besides – sprayed over his letter like a Jackson Pollock. Tellingly the father, an entomologist, remembered them all. It's one thing to sit in a seminar room in clinical training and reel off the latest more or less measured critiques of psychiatric diagnoses: 'oppressive speech acts', 'incestuous bedfellows of corporate psychopharma', 'the last inchoate utterings of a terminal profession'; it's another thing to meet the person they attempt to explain. That he was just a boy made it more unsettling.

Then he didn't come.

'Why are *you* here then?'

'Yes, why are we here?' She looks to her husband: 'Tell him ... *tell him*,' she says, lips whitening. 'You can't, can you?'

When parents bring in an extremely disturbed child the rule is not to waste time assessing them; that self-evidently it's the child who is 'abnormal'. But the absence of the boy meant I had no real choice.

'Well, he's messy. He gets into scrapes. Doesn't know what's good for him, doesn't know when to stop either ...' words thrown into a well, waiting: but there is no bottom. Momentarily the smile pasted on Milner's face peels at the corners. Once darkly handsome perhaps, Cossack-like, now pudgy, a little collapsed, on his way to lost.

'We bought him a Hornby train set not long ago, hoping it might appeal to his imagination.'

'*You* hoped that, *you* bought that ...'

'*I* bought it.'

And as he tells me about the hours he spent happily doing the same with his father, she subtitles him from her side of the sofa – eye-rolling, jaw-clenching, sublingual tutting, a long withering shake of the head – in case I miss how irrelevant, how idiotic all these words are.

'What sort of scrapes?' I interrupt.

'His uniform is always a mess, ink pens burst in his mouth, he's late for class.'

'He put toadstools in his brother's porridge.'

'He thought they were mushrooms.'

'He nearly broke his neck jumping out of the bedroom window.'

'It was only his foot. It was a stunt to win friends; he's a daredevil.'

'Daredevil? He's *the* Devil.'

'Just showing off.'

'Yes, he just showed himself off to the girl two doors down: she's only seven.'

They have taken over. I make notes to make it look like I have a function here, make myself feel like I still exist. I wonder what lengths the boy must go to, to feel the same in their company.

There is a printout of relevant hospital records attached to the GP's letter; the boy's brother, then aged four, had been hospitalised as a precaution, having ingested some mushrooms of unknown origin. He had been hospitalised twice before: once having been force-fed a drawing pin; another time having been struck on the head by a lead pendulum, the result of being locked inside a grandfather clock. My patient's admissions to hospital included for the pinning of a fractured ankle, aged seven, having fallen sixteen feet from his bedroom window. There were several presentations to A&E in the last few months. Five weeks before the appointment, he was in hospital for a relatively minor operation; on the day he was due for discharge he had deliberately thrown himself out of bed, reopening the surgical wound and prolonging his stay by another week. Two weeks after that – ten days before we were due to meet – the boy was back in hospital overnight for reasons that weren't documented.

Tellingly the father had *forgotten* them all. One teacher in clinical training, a psychoanalytically inclined Texan, had told us to always be on guard for denial in our patients: 'There's always an elephant in the room; go make yourselves elephant hunters.' (We had laughed at the self-importance, at the different layers of inappropriateness.) 'Messy uniform', 'scrapes': they seemed woefully mis-attuned, euphemistic at the very least, bunny rabbits in a room of elephants. As Milner offered them up as details of his son's delinquency, he looked dissoci-ated and exposed, that same weak smile barely reaching a foot into the world. I remember thinking how there were no discernible traces of what I read about the boy in the face of this stiff, prudish man and vice versa. Therein lay part of the problem, no doubt.

'He's got a high pain threshold,' he offered a second time.

'No he hasn't. There's no such thing, is there?' She's asking me.

'Technically there is, isn't there?' He's asking me.

'Is there?' She's asking me again, the referee.

I keep on scribbling, a stream of half-cooked formulations, more pretentious than insightful; like

B doesn't attend preliminary assessment, but it is the parents who are really absent.

And

They are complete in their misery ... Complete but dependent – if B didn't exist they would have had to invent him.

Really I am just avoiding two pairs of eyes looking at me, waiting.

'Well, it's complicated,' I say.

And, as I had recently learned in the doctoral programme, one can't deny that pain *is* complicated. In Darwinian terms it's closely allied with danger, the primary organising principle in human behaviour. Its significance is reflected by its complexity, having more phenomenal nuance than any other sensory percept, with parameters that are only just being tentatively sketched in the Literature (Crittenden 2008). Thresholds only make sense in the context of tolerance, magnitudes, sensitivities, pain histories, expectations, temperaments, predispositions. Then there's different potential motivations; one can't assume harm-avoidance, when novelty-seeking or reward-driven actions might just as easily govern behaviour. And that's not even touching on the causes of our pain which may be limitless, anything under the sun: studies suggest the possible stimuli which elicit pain appear to be unique to each individual. Mr and Mrs Milner's fight is part of a mass brawl in clinical practice, the specialties of neurology, psychiatry, anaesthesiology, orthopaedics, psychology fighting to avoid it, resenting its refusal to yield to their limiting beliefs, while each still lays claim to its intellectual provenance. Pain – the leper, the patient everybody cares about and nobody wants to touch.

'He's a thief. He's stolen from us both for years, and our neighbours, and shops,' she says. 'Fags, pens, stickers, chocolate bars, crisps, he eats huge amounts of crisps, eats them like a shark attacking a speedboat. And living things – insects, worms, spiders, frogs, a sparrow once, still shaking – why would he do that?' She's asking me.

To create 'a hideous spectacle to enforce your charity' ... may also fit with high pain threshold ...

Still writing rather than speaking. Really I lacked the confidence of my convictions. What I meant was that a distorted parent–child triangle in which the child is unable to see and understand factors that motivate adult behaviour – like unacknowledged, entrenched marital conflict – may result in theatrical acts of compulsive internalising (sparrows?!) or spectacular suicide-play (window jumps) as a way of attempting to elicit more legible parental behaviour. (Through it I can hear the boy saying, 'Ta-dah! ... Here I am! ... Over here! ... Love me ... Despise me even, but notice me!') Equally it may induce a dulling of awareness in the form of extreme pain tolerance.

'It's hard to be sure ...' is all I do say.

She tells me how he would come home from school, seal himself into the living room, and begin a ritualised relocation of any object that might interrupt the field of vision between his chair and the television: wooden ducks, African elephants, a one-legged Lladró crane, a faux polar-bear rug, all had to migrate across the living room, to be piled up with dozens of books, records, assorted early-industrial brass objects, out of sight behind the sofa: 'Same thing every day, just so that he can watch his shows. Why would he do that?'

An anxiety-driven re-enactment of sorts ... The house was already empty and now B was making it bare ... only he can never remove all the angst.

She's stopped waiting for me to answer.

'Nobody dares go in. Every so often I'll wedge open the door and throw in a multipack of Monster Munch or Penguins like a zookeeper.'

Her voice thickens with grief. 'And I just stand there listening to him eating ... I let it happen ... because as long as he's in there it means his younger brother is safe and there'll be peace ...'

She's exhausted herself into silence. Nobody speaks. Then,

'That's why I bought the train set; it meant that we could have our living room back,' Milner offers.

'AAAAAAGHHHHHHHH,' she screams, a cat's desperate yowl. 'CHRIST A-FUCKING-LIVE ...'

She will scratch his eyes out ...

Her pain was authentic. But Mrs Milner wasn't being entirely honest at that first meeting. I would later find out that she'd been having an affair with a psychologist forty years her senior who taught her weekly evening class.

Under her professor's personal instruction she had introduced behaviour-management plans at home in the form of token economies – credits for good behaviour, debits for bad – adapted from the Literature on children with learning difficulties. Only the boy wasn't learning-disabled, but quick, like a thief. And she couldn't master her brief: there were always glitches in the rules, loopholes in the contracts, which he would feel out before the ink had dried. He made a small killing each week to spend on 'sours' and cigarettes and later glue, butane, Southern Comfort. ('All the other kids want to be Batman or Superman; he wants to be Sue Ellen Ewing.') Tragically for Mrs Milner, the boyfriend died in the middle of the intervention (always a risk when there is half a lifetime between you), so then she had to cope with him – and her grief – on her own. The programmes were abandoned. Milner knew all of this and did nothing and then one afternoon he bought the train set, three months before his son's birthday, to piss on his own lamp post again, his gift really a weapon in a long, shapeless fight that started before they were married.

Moments pass. I don't dare make a note. She has her hands clasped to her ears, framing her bereft face, her husband less than three feet away, still smiling.

Like a diptych titled 'Together' ... How much of all this is for my benefit?

Looking back it seems strange that Milner started telling me about the toy and his purpose and then repurposing of it, given what the boy would go on to do with it. There's a particular amnesia at work here, something about people retelling the past – prompted by a momentary rush of excitement – whilst forgetting where it leads, and what where it leads might say about them. Friends and colleagues, noting I practise this on a regular basis, jokingly call it Benjamin's syndrome, my

sole contribution to the diagnostic canon. But it's not specifi-
cally mine; we all tell stories against ourselves.

'We're running out of time today,' I lie. We are barely
halfway through.

The boy didn't show for his next appointment. Strange to say
I felt more disappointment than relief when Mrs Milner came
without him. She had brought his photo instead. A featureless
moonish face, bossing forehead, crooked overcrowded teeth,
otherwise unremarkable. She wanted me to *see* something –
monstrous derangement? Exceptional pain? The possibility of
transformation? But it was just a photograph. Looking into
his small, dull, grey eyes, aware of hers fixing mine – hanging
on whatever I might say next – I was struck only by how
empty I felt.

Given permission by my supervisor I was allowed to see
her over the following weeks, with the rationale that I might
positively influence the child's parenting. And I thought how
despite his absence the son was finding different ways of
contacting me, with behaviour that ensured his parents' presence, a gateway to marriage guidance perhaps, and when that
failed, support for his mother.

Given the space, Mrs Milner talked in torrential, breathless
detail. Always the same question: 'Why, why is he like this?'
Asking, answering for herself, then demolishing her answer
and asking again. Her stories snaked towards intimacy and
closeness then suddenly jagged back. She told me how while
she was pregnant she had unusually powerful fantasies of how
she and her boy would be together in the first years of his
life, a substitute for a marriage that was already dead in the
water; in her imagination she'd made her baby her partner.
And the sweetest of ironies; the baby was due on her wedding
anniversary. But because the labour lasted seventy hours ('He
wants to ask *me* about high fucking pain thresholds!') it meant

that they shared the same birthday; they couldn't be closer. And for a moment she is twenty-five again, marinating in the loving opioids of new motherhood. The next moment her face has curdled.

'He fought tooth and nail to stay inside, why would he do that? Why?'

Foetuses have no teeth.

Eventually they have to use forceps and when they finally pull him out, his huge head – the cause of the problem – has been squashed and bruised, so it looks like a giant black dinosaur egg. Some birthday present: the creature from the black lagoon.

And his face, his first face, looking up at her, asking something as she asks me. And we (she and now I) don't know what is meant, can't placate these faces …

'And what about the face before that?' she wonders.

She means the prenatal one while he was still inside her. She will never know, but she still imagines it forming, registering, layer by layer, the despairing muffled rhythms, the bleak cadences of the Cold War taking place around him.

'No wonder's he's like he is … He didn't even have a name. *He* wanted Ben and I never really did.'

Weeping, she tells me that though she would hold him and kiss him passionately, call him one name and when that didn't 'feel' right, another, and then another, until no name was right, and Ben would have to do, it was already too late, all that uncertainty – that absence of *intuition* – had done its damage.

A decade on and they hardly ever speak. Apart from those rare nights, three times a year at most, when he wakes her, leads her into his bedroom away from the father, and barely conscious himself – a kind of sleep-talking – tells her every little thing lying on his heart. The next morning, distance resumed, he remembers nothing, and she is left with what he told her. So she does the same to me. As though at his behest it is my job to convince her that, carnage apart, the rare summons in the dead of night was evidence that the filaments of connection remained between them; meant their bond, however fragile, might one day strengthen.

Many years later I bumped into her outside the train station in Leeds where I was attending a conference. It was nice to see her again. She smiled and I remembered noticing – it can't have been for the first time – she had the same crooked teeth as he did. Had her parents been similarly preoccupied, missed dental check-ups, and God knows what other aspects of her care? The kind of possible diagnostic sign I had picked up over the years, but wouldn't have occurred to me on such an early placement.

Ben had just turned thirty.

'What's he up to?' I ask.

'He's working at a homeless charity. He helped set it up.'

'Looks like it's a problem here.' On my short walk down a recently vajazzled Victorian arcade (every third shop was for hair removal or tanning) I had been asked to buy a *Big Issue* four times.

'It's a problem here, especially for him. He looks like one of them, smells like them too, and he's always asking for money which he gives straight to them.' She was still the same mix of hard and soft. It sounded like he still 'enforced her charity' – her own bespoke Bedlam Beggar.

'Can I ask you something quickly, Doctor?' and still the same impulse to say more.

Always somewhat relaxed around clinical boundaries, I agreed and we went for a coffee in the half-hour I had before my train left. Back then I had referred her on to a psychotherapist who had been 'less than useless' (she didn't elaborate) but somehow the experience had given her the fuel to finally leave her husband, her therapist, men in general. Her children apart, she had close female friends, enjoyed her work, had found a home for herself with a local Buddhist group.

She spoke about him; the chaos of his teens and early twenties, scarcely believable, scarcely bearable, was convinced he would end up dead or insane, still worried about it from time to time.

'He lives like he's balancing gelignite on top of this head. Or I do. He doesn't even know it's there.'

Some things can't change. As before, she filled every inch of the space between us with her concern, enslaved by the same rhythms of anxiety and remorse.

'If only I knew then what I know now, I would have been better at it.'

'I'm sorry he's had such a sad, difficult life.'

She told me how she read psychology and self-help books non-stop. Did I know that there are three types of child: those who are well-enough cared for, those who are eternally lost (caused by a 'fundamental disturbance of basic structuring', she remembers), and then those in between, the ones who might go either way?

'Which one is he?' Her limitless mind always snagged on a single point.

Really, three were only two, I wanted to tell her.

'There's still time,' I lied. With the perinatal trauma and his early attachment history he was always likely to have serious problems. Likely, but not certain. 'I never met him, Mrs Milner.'

'You didn't.'

'I should go.'

'Funny he never wanted to come, given how much psychiatry and therapy and rehab he's had, on top of AA and NA and CA and SLAA and I don't-know-what-A ... Only *they* don't know him as Ben; he's Junction Box there.'

★

· 26 ·

'Together', a diptych, with me in it, only out of view. Together, but not for much longer. Milner left the family home a few weeks later and took his eldest son with him, which wouldn't be good for either of them. Together, but not for much longer – we had seven minutes of the hour remaining and I was still waiting to hear what had really brought them here now.

I didn't have to wait long. Her scream released her, a runaway train, describing how her husband had carelessly set up a section of track on a large wooden board on his son's bedroom floor, provided brief instructions about how the engine wouldn't move unless the circuit was complete, then left him to it. She noted with thick, sarcastic delight how the vehicle, a beautiful dark blue Mallard replica, never left its box. Her son had configured the track in the most basic, unimaginative loop possible. The level crossing with its operational traffic lights never made it onto the plywood stage, nor did the realistic moorland hillock with built-in tunnel. Amputated workmen and village types were strewn across the floor like they'd stepped on IEDs or drunk the afternoon away, a headless mayor rode on a pig, a Barbie doll (recommended by her psychologist boyfriend to promote the boy's femininity – Ha!) straddled the track like a porn star. If you were a play therapist you would tick every box on the risk pro forma.

But these were red herrings; it turns out his imagination was elsewhere.

She described how he came straight home from school and, grabbing family packs of Blue Ribands and Quavers, headed up to his room. He had a sticker on the door now: 'Nuclear-Free Zone – Keep Out!' (The psychologist had also encouraged her to allow limited 'alone time' in his bedroom to promote his capacity for solitude – Ha again! Where would we be without professional psychology?) He'd declined the offer to have friends round with their engines. For the next few days Milner would get home from work and, seeing the living room in order, the younger son quietly watching cartoons on television, give her the lightest of smug looks.

As is often the case, the research began quite accidentally. He had been idly messing with the track when a small shock thrilled his arm. The junction box had a dial which controlled the train's speed: the higher it went, the greater the shock. He plugged in each finger in succession, carefully testing it, noting the different characteristics.

Seeking a physical correlate for emotional disturbance?

There was always a point where pleasure and pain were balanced perfectly, but it was momentary and different for each finger. It changed when he sucked them first, or when he wore his verruca swimming socks.

Creating a reliable sensory environment as a substitute for mother-as-environment?

So many variables, the balancing point always in motion, like the subtle differences in how long the sensation takes to travel up his arm. Where does it go after that? It feels like it's heading for his heart.

May compromise future intimate relationships?

The main thing: to be as systematic as possible. Systematic and dedicated, like a nineteenth-century toxicologist experimenting on his own body. Had he known, his scientist father would have been so proud.

May also be a way of attacking his own sensitivity?

His project intuitively leads him to the more intimate dimensions of sensory experience, to lie his giant half-dressed body across the track, like Gulliver, or a ravaged, outsized heroine from the silent capers featuring Harold Lloyd that are on before the news: 'Heyulp! Heyulp!'

Names ... Name-changing ... She could never find a name to fit him properly ... So Junction Box creates his own, imposing himself on his absent parents in intensely imagined ways. He takes their dead gift, the weapon they used against one another, and makes it into his own — something evocative, erotic, dangerous, shocking, so much so it becomes his Proper Name.

I would make this last note on the train home from Leeds. For all my attempts to interpret his behaviour, I really hadn't understood the young boy's project at all until then. Over the years my familiarity with pain deepened, and I discovered for myself that at high levels of arousal in those with high thresholds, there can be a sudden counter-intuitive reversal of charge so that pain is experienced as pleasure.

'I think it was the first time he'd really felt himself.' She pauses a moment. 'I was making his tea. The Findus Pancakes were still half frozen in the oven. There was the smell of something burning. I heard the young one screaming. First thing I thought was he's killed him this time. Honestly, even in that moment before I knew what was going on, part of me was actually relieved that something had happened, that this train idea was full of shit. I found James at the bottom of the stairs squealing that his hair was on fire. But his hair was fine, there was nothing the matter with him.'

Pain can be experienced empathically, by pure observation, as well as by direct physical or emotional stimuli: all three produce activity in the same brain structures.

'And then I see him, standing at the top of the stairs, school shorts round his ankles, laughing and crying at the same time ... He's cupping his charred little thingamajig in his hands, saying, "Sorry, Mummy, I'm sorry."'

And as she cries quietly into a tissue I've provided, I turn to the father who has sunken further into his sheepskin, as though digested by it, still smiling blankly in my direction, and I think how much he looks like the father in *The Railway Children*, black hair, bearded, blue-eyed, the one who goes away, wronged, imprisoned, who goes away until he comes back.

Lucy

'Could you hold up your left hand?'

'My left hand?'

'Yes, hold it up, please.'

She gets it right that time; that's two out of three.

'Have you had difficulties driving?'

'No, I don't know ... should I?'

'Show me how you would use a pair of scissors.'

'Scissors for what?'

'Cutting hair.'

'I go to the hairdresser's.'

'Cutting anything.'

'You mean like this?' She snips the air with her second and third fingers. 'Is this what you want?'

It is not what I want; it is how scissors are symbolised, not how they are used. I make a mental note; writing down would only worry her more.

'Wave please.'

'What do you mean?'

'Just wave goodbye.'

Tentatively she raises her hand and waves as though trying out a new prosthetic limb.

'Like this ... ? Or is that hello?'

She had Alzheimer's or vascular dementia or corticobasal degeneration or nothing. She had the same name as my mother, was close to her in age, wore the same floral dress that mum might wear. There was a long day ahead. After me there would be bloods, a spinal tap, magnetic imaging, electro-encephalography, finishing with the consultant neurologist who would make a decision one way or another.

She was not doing well and she knew it. The lines on her tired, ageing face gathered like a storm-map. She was sweating, on a November morning in a clinic room that was so cold I could see my breath. Even the thin, creased lobes of her ears beaded with perspiration. Her self-report was confused. She had either mistaken her neighbour's house for her own, or she was just watering the plants as a favour. She flooded the kitchen answering a cold call while washing up, or the washing machine wasn't working properly. It took her fifteen minutes to find her way back from the Ladies. She'd been twice already, we were less than an hour in. There was water everywhere: her stories leaked, sweated, flooded, wouldn't be dammed. She was like and not like my mother.

Names would stick on her tongue like peanut butter. A 'whisk' was a 'whick', then a 'witch', then a 'whipper'. She couldn't recall the prime minister before Cameron.

'Is it Blair ... ?' (Trying to read me for clues.) 'It's not Blair ... It's the sulky one with jowels ... I don't know; they're all as bad as each other.'

Most of our faculties diminish with age; otherwise healthy brains atrophy. On my last visit home I was struck by how frequently she used 'whatsit' and 'thingamajig'. It could be normal ageing, just about. We had become a little closer these last few years but it would always be fragile.

'Oh, for God's sake, Lucy, come on! He was so *forgettable* …'

While her version of what was happening shuttled between catastrophe and denial, within the same sentence, I bobbed on the feeling of familiarity. A mother and son mirroring each other's confusion.

She spoke of the world she came from. In the previous eighteen months she'd retired from a long career helping others, then suddenly lost her husband of forty-three years to pancreatic cancer, had to give up a plan (fifteen years in the making, including intermediate Spanish classes and two recces) to spend the first few years of their retirement working with indigenous children in Bolivia. Her own children had long since left London in search of affordable housing. The majority of her friends were infirm or dead. She had recently been diagnosed with type 2 diabetes, to go with irritable bowel, chronic pain, shingles, and reduced mobility associated with a failing hip. In those liquid eyes, trembling now with fear, kisses for all she loved could still be seen. With disarming, baleful honesty she described how despite frequent joys, her life had always been a struggle to keep her mood afloat. But nothing had prepared her for the assaults of the last few months as she waited for her 'urgent' outpatient appointment; insomnia, panic attacks, loss of appetite, agoraphobia. All this in a woman whose other shelved retirement plans included potholing and scooting around Zone 3 on a Vespa.

These factors could not be ignored. Diagnostically it was possible that they preoccupied her to the extent that profound distractibility was causing her symptoms. Added to which her brain and its pathways had, she believed, long been grooved by an attritional emotional climate – *that* had to affect her thinking. But that her life was blighted and that her brain in some way reflected this didn't protect her from having an additional curse in the form of Alzheimer's or another disease: it may even have predisposed her.

As full of grief as age: Lucy and my mother.

'It's a sss ... ssst ... er ... you use them for listening.'

I am pointing to a stethoscope.

'... Steh ... no, it's no good. Can I tell you something?'

'Please do.' But I am not really there: I am looking out of the window at a thin sock of smoke which rises from the incinerator chimney, thinking about my mother who recently changed her mind and asked to be cremated.

'I always wanted to be a French ... no ... I always wanted ... um ... er ... French ...'

She is confused, I am distracted. The sweat patches are slowly moving across her dress, a live map of drifting continents, like someone drowning in slow motion.

'I wanted a French thingamy ... er ... I wanted French letters ...'

I was part of what was happening, an active ingredient. The Literature indicates that if the doctor disengages at a critical point in his patients' storytelling, her speech may become dysfluent or semantically confused. So is it me or her causing this confusion? It's possible her differential included a primary progressive aphasia (a variant of Alzheimer's that begins with disruptions to language). Equally the hesitations and mistakes may be no more than the effects of our interaction, clots in the blood that connected us. There is no view from outside, no possible third-person perspective that isn't just 'me' and 'you' making our way uncertainly, impressionistically.

She should be seen by someone other than me. Concentrate: eye contact, thoughtful pauses, patience, encouragement, mirroring, reassurance, gentle speech, reassurance without misleading: rituals, like puja to a god, like love, love in fact, which might somehow guarantee that which passes between us.

'Alasdair ... ? Are you there?'

Only my mother calls me Alasdair.

'Yes. Please be patient,' I tell her.

★

Later, as I made my way down the huge open-air lido, the morning clinic over, the sun cast my shadow on the pool floor. I watched my silhouette reaching out in long strokes to get a little more water, a little more speed. To catch myself? To escape? What could this striving mean when my real competitiveness as a swimmer was years behind me? Four feet below, my mind is a murky, shapeless blank that can't be read. Sometimes when I come to the moment of diagnosis I have the feeling that it is happening without me. I've stopped hearing, stopped being heard, retreated beyond recall. Whatever she or anyone else might say to try and change the outcome, 'decisions', if they can be called that – the layers that form them may be so thin and fragile they barely have substance – have been made.

Back in the office the Dictaphone poised an inch below my mouth, I was still hesitating. The dress was just like hers. Features of her performance were borderline, well below expected levels. My mother is too thin. Her speech running out. If no diagnosis was made she may lose valuable months on a neurodegenerative retardant. I wasn't sure. It would be natural if after years of being inundated with terrible trauma and terminal disease, I have gone the other way, become too quick to normalise pathological symptoms, especially where friends and family are concerned.

I pressed 'record' and started to fill the silence:

'Dear Dr R, thank you for referring this amiable though anxious, well-dressed sixty-nine-year-old right-handed lady ...'

Nine months later and, the name apart, any resemblance to her had gone. My mother had remained more or less the same, appearing to hold her own for now. But this Lucy was in free fall: her face had lost tone, her hair had thinned, her gait rocked like someone ten years older, the buttons on her jacket were not fastened symmetrically.

A decision had been made.

On the afternoon of the same day I originally assessed her, neuro-imaging had found 'hypo-perfusion in the frontal poles' – reduced blood supply to that part of the brain – a possible early biomarker for neurodegenerative change. Although the radiologist reported this as inconclusive, Dr R, the neurologist, thought otherwise. He reviewed my carefully hedged report – I had recommended a year-long deferral – and of his own accord thought the different lines of investigation suggestive enough of early dementia to make a diagnosis.

He shared his thoughts with her.

'It's not good news I'm afraid ...'

I imagine he would have told her about the basic mechanism of the disease, symptoms, prognosis, the limited treatment options. Likely the dementia nurse specialist would have met with her and discussed available support, possible compensations, dietary changes, etc. At home friends would no doubt have told her their stories: about a young man in his sixties two doors down ('It was because he retired') or a woman from the gym ('Her eyes turned milky overnight'); blunt, half-concocted vignettes whose content and rhythms had been absorbed unquestioningly from the *Daily Mail*, Jeremy Kyle or whacked-out online forums. In these ways a new story is instantly begun, begun and all but finished.

Nine months on and the anxiety that charged our first meeting had hardened: water turning to stone. The reassessment indicated a significant downgrading in the domains of language and memory in keeping with her new diagnosis.

Later that same morning when she was re-scanned, the hypo-perfusion had disappeared. In his afternoon clinic the same neurologist explained that the original MRI finding had probably been caused by a transient phenomenon like reactive depression. Suffering, you might call it. She does not have dementia. Not now, not then. A major error had been made. It's unusual, very unusual, but possible. In her case the MRI was a will-o'-the-wisp. Given the brilliant emulsified detail, the brain-like beauty of those images, it can be hard to keep in mind that they are not the brain itself but an analogy penned in nuclear physics; ambivalent, layered with artefacts, esoteric like the Kabbalah, readable and therefore mis-readable. Maybe the neurologist was distracted that day, worried about *his* mother, his wife, his lover, the repayments on his second home. Unlike some of his colleagues I have always found him to be honest, well-meaning, dedicated. I imagine he apologised sincerely, was upset with himself for more than just a moment. In the fifteen minutes allocated to her he would have no doubt explained how her language and memory symptoms were a mix of 'feeling a little blue' and 'getting on a bit', then discharged her. She would have gone home to the same friends and neighbours who would be re-confirmed in what they always knew: 'Typical bloody doctors.'

Six months after that, following an angry re-referral from her GP, she is back in my office reporting further deterioration and significant functional signs of moderate-severe dementia. Or rather a neighbour – the one person who hasn't drifted away over these last few months – is, while Lucy looks like she's been bolted to her chair, a horrified ventriloquist-less dummy, her jaw mouthing silently, tears streaming down her face, unable to make the opening sound of any word. Every muscle and sinew around her tiny skeleton is visible and painfully contracted, as though flayed. Her scalp, poking through thin lank hair, is corrugated with a yellow crust, the skin on her neck is blistering, her teeth a deadly grey. She is wearing a brand-new coat, the label still attached, over a faeces-stained dress. The decision had been made over a year ago by the experts. Back then it fitted with her experience of herself in ways they couldn't possibly imagine. But now *she* has imagined it more fully, grown into it – her Electra, her Ophelia, from a woman who all her life had been content to remain backstage – so deeply she cannot be called back. Try as we might to invoke another possibility – good health – there is no longer room for it: the audience has taken over the theatre, the real transformed into a ghost, or a shadow.

She is passed on to Psychiatry.

I have a personal investment in this strain of misinformation, what is known as diagnostic foreshadowing. In my late twenties I went to see a middle-aged male psychiatrist, who had been recommended by a friend. Though I often minimise it these days, I was terrified. A fragility whose roots disappeared into earliest childhood had been awakened by a sequence of ordinary blows: the break-up of my first serious relationship, the slow failure of a career in the arts, the dawning sense – common to that age – that life comes to a point whether I want it to or not. As the weeks passed the acuity only sharpened, the powerful magnetism of collapse attracting otherwise unrelated fears and it became obvious that I lacked the safety net of the psychologically robust, a break-fall that I had assumed was part of my inheritance, part of the deal of being white, educated, bourgeois (wasn't well-being written somewhere in the small print?). And then one late summer evening in 1999, superficially like any other, I had to drag myself off the platform at Tottenham Court Road, escort myself up the escalator and out of the station, and frogmarch all the way home, so taken was I with the idea I might jump. What I remember of the assessment was the sense that in my pain and befuddlement I was unguarded, careless; desperate to be heard, to do justice to my feelings with both accuracy and conviction. Though obviously pressed for time the doctor was friendly enough; he *seemed* to believe me. Every so often he would make noisy, exuberant ticks with his pen, leaving me to imagine what was being confirmed. Then it was over.

A few days later at a follow-up appointment with my GP I was told I met criteria for a major psychiatric disorder; difficult, opaque, disorganised feelings had been checklisted, organised and medicalised; that is, made simpler. I had been believed, I had been understood, but in his terms; as though understanding was something done to you, without you. The psychiatrist prescribed a combination of drugs that I was to take indefinitely. I never took the prescription to a chemist. I never saw the psychiatrist again, but for years afterwards I struggled not to understand rogue moods or unconfined thoughts in light of his explanation, still can on a bad day. The label – wrong though it was – stalked me, bending otherwise innocuous states and behaviours to itself. And a further dimension of the peculiarly self-fulfilling, malignant energy that such (mis)diagnoses carry: I imagined everyone else understood me in the same way. Tick, tick, tick. At times it has required huge amounts of defiance to keep his name for me at bay, which carries its own costs and distortions. The experience is part of the reason I do what I do now. And never far away, the temptation to jump, accept it – like making the best of an arranged marriage, so that over time you forget that it chose you – live up to it even, for there were perks too.

Me, Lucy, my mother.

Misdiagnosis is towards one end of a spectrum that includes slips of the tongue, exaggerations, more or less conscious distortions, partial truths, biases, screaming lies. I organised a small clinical research group to investigate the parameters of deliberate misinformation; how easily it might take hold in the hearts and minds of the normal, not just the obviously fragile and infirm, and crucially whether its force might be sufficient to make nothing into something diagnostically.

In our first study we took a group of healthy volunteers, young medical students for the most part, and got them to rate themselves on a symptom checklist of potential physical, cognitive and emotional difficulties. Next they carried out a few simple memory and attention tasks using instruments that are not uncommon in diagnostic neuropsychology. Then they entered a clinic room where a senior consultant (actually a mature psychology graduate with good acting chops – white teeth, expensive suit, deep unfaltering voice) interviewed them about their reported symptoms. Unbeknown to them a number of their symptom ratings had been changed according to a predetermined template. So, for example, 'no reported memory problems' was changed to 'some reported memory problems', 'mild' anxiety became 'moderate' anxiety, and so forth. On that basis the fake doctor questioned them on the details of problems they hadn't reported.

People express beliefs or preferences or accept descriptions relating to themselves without really knowing what they're talking about. That they do so means these beliefs can be manipulated, even when they may pertain to the most intimate aspects of who they are. Our deepest sense of ourselves may be blind spots that others may fill in without our knowing. If during the interview the patients piped up and said, as was actually the case, that they hadn't indicated memory problems, then they were not 'change blind'. But to speak out, to say 'all is well', is to contradict the authority of the doctor, whose power – stretching back for thousands of years, across great civilisations – has now gathered in this satanically handsome mimic's conviction-filled eyes. And at the same time it is to override that inchoate sense of something always being wrong, which sits more or less buried in every one of us.

Nearly 90% of participants were 'change blind', appearing to accept the misinformation about themselves.

The lie entrained, our consultant encourages his 'patient' to elaborate it further:

'So, Miss Atkinson, you say here you have difficulties with your memory; could you give me some concrete examples?'

And anyone can find recent examples of absent-mindedness if they're looking for them.

'Oh, I see, you forgot *all* your shopping, and you didn't realise your error for twenty minutes, *twenty* … ?'

Implied incredulity, seasoned with lightly pathologising concern, '… Yes that does sound a little worrying,' as he makes a tick of his own with an expensive-looking fountain pen.

The line of questions is pursued until it finds what it wants. The patient sees a mind making itself up in front of her, in spite of her.

'And just how often is this sort of thing happening? Try to be as precise as you can, it's very important. I see … And has anyone else noticed; your boyfriend? Family? They have? How about friends? Oh, I see.'

He *sees*; the grand diviner, the oracle. Miss Atkinson is nudged, with rising fear, into telling a story which she believes is congruous with the expectations of the doctor, the expert, even when the story is not hers.

After the interview the subjects completed different versions of the memory and attention tests they had taken before the interview. On more sensitive tests they performed significantly worse afterwards. The margins of these changes, though small, were sufficient to meet diagnostic criteria for the early stages of a neurodegenerative condition. (Participants were given sensitive debriefings by a trained counsellor at a later date.) We concluded that poorer performance was associated with change-blindness and reinforced false beliefs about poor self-ratings on the targeted cognitive domains. In other words, given the right conditions almost all of us need only the slightest encouragement to behave as though we have lost the most fundamental content of our minds. The results are provisional, the sample size is small and biased, but they at least point to various potentially subjective – intersubjective, more accurately – aspects of the diagnostic process.

There are mistakes; they are inevitable. Then there are the people who make the mistakes, how they make them, with what conviction: how important is being right to who they are. It's not just a pill, the doctor himself is a placebo, or its negative in this case. Then there are the individual differences of those who are mistook: their fragility, their suggestibility, their neuroses – how important being well or ill is to who they are. And overarching all, though often scarcely visible, the medical *mise en scène*: bone-white walls, blue scrubs, signs that say 'Nuclear Medicine', a tendon hammer, the stale smell of lymph in corridors, other assorted props which somehow aggregate with astonishing cultural force; the background without which the doctor and patient and their love affair would not be legible.

It was as though she had been bound fast to the patient's chair, but there was no tether other than her belief. She was a fast-forward of what might have happened to my mother but didn't. And it hadn't happened to her either. We identify with symptoms when they are never us. We identify with symptoms even when they are not there. The curtain descends at the moment of diagnosis, leaving her cast as this horrible gorgon. But the genius was her own: like a photo taken the instant of a bomb-blast, unstageable by anyone except the explodress herself. She looked like she was drowning in herself; a body thrown overboard by her own hand, sinking like a stone. *In the bleak midwinter, frosty wind made moan; Earth stood hard as iron, water like a stone*: my mother's favourite hymn.

Diagnosis is not the terminus, though it might become so by an act of will. The conversation continues long after we leave the clinic room, a million new synapses in a narrative where provisional truths or falsehoods may take hold again. We are required to find ways of defending ourselves against certainty and uncertainty alike, and know that even at our most conscientious, our most conscious, important aspects of our decisions are beyond us, wrong possibly, made as though by a shadow who overtook us without our knowing.

Michael

The two of them had gone to Chamonix for the week. It was the first time either could remember being on holiday without the others. Michael felt stronger than ever. He cycled twelve miles each way to work, not so much a commute as a series of jackknife anaerobic sprints between lights, racing against the increasing number of furious, shark-faced cycle-commuters on their five-kilogram bikes. He swam at lunchtime, when he could, which increasingly was most of the time, did burpees in the work lift if it wasn't too crowded, made himself 'plank' during conference calls. Last year he haggled four months' worth of leave to join a friend – ex-Gurkha, lost a leg stepping on an IED – who was retracing a POW escapee's journey from Siberia to Darjeeling.

Luke's birthday would fall in the middle of the week. Michael had bought him a new jumpsuit as a present: the Rebel Freebird, for advanced basers, commemorating that once-in-a-lifetime moment where the son turns half his father's age. It was natural that they should race.

So: a blood-red Phoenix, each limb ending in a burst of flames, versus a lime-green Flying Squirrel whose tail (surprise,

surprise) will spell 'Happy Birthday' when aloft. Even Michael, who made garishness his signature, conceded they looked a little ridiculous, perched in nylon wing-suits on a blustery ledge below the Aiguille du Midi, 3,000 metres above the valley floor with plenty of lethal obstacles between them and it to avoid a boring descent, waiting for that perfect moment. The bird was three inches taller and several kilos heavier than the rodent and he'd spent most of last year jumping in the Colorado Rockies, including the sort of 'through the keyhole' moves that made even Michael's arse twitch. Such was his instinct that when the moment came he had already launched. By the time the Squirrel's older reflexes had responded, a headwind had replaced the lull and the Phoenix was already a 150-mph blazing speck in the distance.

Michael has no memory of the view he enjoyed, or anything else that happened in the month before his collision, and unlike his son he was anti-GoPro (for him the world divided into two mutually exclusive groups; people who took pictures and people who starred in them) so there is no black box – human or electronic – to review the 'decision' he made to fly towards the *péage*. Witnesses spoke of a struggle to engage the chute, then straying towards a lorry parked in a layby – it seemed 'magnetic' in its attractiveness – next to a sign welcoming drivers to the Alps in English, French and German, each letter thirty-centimetres-high Helvetica. He clipped the vehicle's wing mirror with his head at approximately 25 kph ('Some welcome'). Even with a helmet the mirror effortlessly removed the vertex of his skull, cupping two cubed inches of cerebrum from the left frontal area. By the time Luke, who had landed half a mile away in the designated area, realised his father was missing, a young French veterinarian had pulled over and was doing her best to staunch the Squirrel's blood with the goose down from her Marmot jacket while her boyfriend called emergency services ('I was the biggest green squirrel she'd ever operated on').

Michael was fifty-eight then, but he looked fifteen years younger in the photographs: lavish shoulder-length black hair, pale blue eyes, the frame of a competitive middle-distance runner, only he found running boring on its own, unless it was running after something. An old college friend had taken a picture of the group on his fifty-fifth birthday outside Bruern, the family's country home in the Peak District. I'm looking at it now in a scrapbook of relevant history his wife Catherine had compiled for my benefit. Next to it another photo, of a graduation party; the same core of people, a few omissions, a few additions, three decades apart. Because of an odd kind of reversal, it takes me longer than it should to match the individuals. For the first time in human history we, the medically super-literate middle classes, are, en masse, in better shape at fifty than we were at twenty-five. Notwithstanding greying, balding, the inexorable drift into chinos, it's hard to match trim and tucked and buff with their pasty, smoking, flaccid younger selves: our brains anticipate change, but in the opposite direction. Except Michael looked exactly the same: hair still improbably black, lineless face, somehow in action even standing with his single malt, not a pound heavier, like an elongated Michael J. Fox, wearing the same *Back to the Future* jeans and white T-shirt. Only the Rolex was different. If the others were acting on an unexamined fear of death masquerading as 'health', or still less tractable fears, then Michael never had such fear to examine apparently, wouldn't know health in the same way a duck doesn't know water.

He was driven to the nearest regional neuroscience centre in Grenoble. The on-call surgeon performed an emergency decompressive craniotomy – sawing out a section of skull to give the damaged brain space to swell into – in a shade under seven hours. He was intubated and kept in a coma on Neuro-ICU for five weeks. Catherine arrived the next morning. She and I would relive these early hours over and over and over again. To this day Michael – beyond wheeling it out as an anecdote for his friends at the White Bear – just isn't interested in what happened or what changed because of it, or can't take an interest, certainly can't fake one. Already the trauma was Catherine's alone: 'Night One' was the first time she would go to bed without a message from him – he even spoke to her daily from Siberia on a satellite phone – the beginning of a separateness that would only deepen.

When he was eventually woken and able to sustain his own breathing he was transferred to an acute neurology ward. Most of the patients were a generation older than him with medical complications secondary to neurodegenerative illness. Michael remained in post-traumatic amnesia for several weeks, meaning he was unable to lay down even the most basic new memories, a function of a neurochemically programmed shut-down of the cortex to facilitate emergency repair. He was dyspraxic, anomic and perseverative. Catherine would tell me how he would apply toothpaste to his eyebrows (dyspraxic), called both the television and the nurse a 'trout' (anomic) and anything else for that matter, until a new word got lodged (say, 'butter') and was used to label all (perseverative). This super-normal, unimaginative, high-end fixer (I still don't really understand what his job was), known for his sense of fun rather than humour, transformed into a compulsive surrealist: Michael, the English Dali, the seventh Python. Fortunately for him there were no orthopaedic injuries of note. Unfortunately for the nursing staff and the general equanimity of the elderly ward, as soon as he was able Michael would stomp through corridors pursued by health care assistants in a state of high agitation, punching, swearing, spitting, groping, frotting, kicking, pissing – behaviour that is collectivised, euphemistically, as 'challenging', and which, his wife assured them in broken French, was quite out of keeping with his character and upbringing, swearing apart.

Given the prevalence of severe traumatic brain injury (TBI) there is scandalously little evidence of the benefits of rehabilitation practice during the acute stage. What we do is primitive: low-stimulation environments, regular multi-sensory orientation, talking softly, a little music; it's *that* technical. Really it is no more than a collective holding of breath, trusting the body will repair itself. (Such is the Cartesian pull that it still feels strange to write 'body' when one is talking of the brain.) It's unimaginably hard on the family, but it's also a difficult time for professionals who are used to urgent, intensive, sophisticated intervention. At this stage any interference can make things worse: antipsychotics will increase the challenging behaviour in this population, as will excessive rehab demands by the treating team when the patient literally may not know his arse ('arm') from his elbow ('whalebone'). The nursing staff are encouraged not to over-personalise responses when the overloaded patient's behaviour becomes challenging. But it's hard for the French team not to take it personally when someone quick and strong and nearly as tall as John Cleese is coming for you holding a weaponised Zimmer above his head, trousers around his ankles, shouting, 'Cock off, frog-fuckers!'

Slowly Michael calms down. The MRI scans look OK, apart from the missing brain: no midline shift, no enlargement of the ventricles, no obvious diffuse axonal injury – telltale signs of structural damage in more severe injuries. Now that the skin has grown back over the skull-less cerebrum there's a clean scoop in the left side of his head as though a larger species has just taken a spoonful from its morning egg. Superficially he has 'bought' what happened to him. He knows where he is – a hospital – without knowing the town or country. He knows what month and year he is in – it's written above his bed in felt tip – though the days are still slippery and the date is pure guesswork.

Used to private jets, Michael is flown back to the UK in an air ambulance where he's transferred to a post-acute rehabilitation unit. After a week of assessments, the multidisciplinary team establishes various goals, supposedly in conjunction with him and his family. The occupational therapist wants Michael to make toast without catastrophe (he had twice tried to insert his tongue in the toaster). The physiotherapist wants him to stop sprinting everywhere, given the huge risk of falling. The speech and language therapist wants him to monitor his swearing and use eye contact to assist turn-taking in conversations:

'Look into my eyes: fuck them apples ... Your go.'

Michael's 'goals' are different. Like creating a dry-ski slope out of an area of flat marshland, or planting a bonsai orchard, or rearing an alpaca herd for the wool (but mainly 'for the comedy'), and, most persistently, cutting a croquet lawn in one of Bruern's sheep fields, a folly for a Christmas summer party on the 27th, straight after the Boxing Day hunt.

'I'll use a Caterpillar to clear the snow, banked up it will make a perfect boundary. We'll use pink and orange balls, stop them getting lost. We'll have Pimm's, summer-fruit coulis, barbecued snow leopard – just kidding ... ordinary leopard.'

The seasonal switching could be chronic temporal confusion. Much more likely it's a rich man's folly. Either way it's hard to find common ground between the patient and his team.

I meet Michael six months later, nine months after his accident, at his North-West London home. Catherine and Luke are there, as are three younger daughters, each of whom look like well-judged impersonations of their father, of what he had been. The five of them will remember that meeting years later because of one thing I tell them:

'Family Taylor will never be the same again.'

They had known this for most of the last nine months, without allowing it to surface. Because equally and oppositely, there is a blindingly wilful assumption that we recover from illness, and with it a fiction that recovery doesn't stop until it's complete. We will become ourselves again whatever that could mean, whatever that could mean for Michael who more than most had a talent for being himself, as though returning to normal were a Newtonian law and not a child's fancy. Michael was up and about, joking, full of himself, and with the new cranioplasty rounding off his skull he *looked* so well, almost complete. So many *trompe l'oeils* underpinned by the greatest deception of them all: life, like each one of its sicknesses, culminates in health. No wonder the fantasy is hard, for some impossible even, to fully close down.

Or it closes down too quickly, because, paradoxically, some deception may well be necessary to any recovery that *is* made. The surgeons followed by the neurologists and the therapists spoke of a magical 'two years', the brain's exceptional plasticity meant spontaneous recovery would be ongoing for that long. One doctor had even spoken about three years. So which one was it? Did I know? Did that mean Catherine should postpone any judgement about what was lost or different or just broken until the full two years had elapsed? The more the professionals enthused about these 'two years', the more she felt it was something for them to hide behind; a gesture of authority and expertise sugared with numerical specificity, but really a mantra, an invocation of the faithful – 'magical', in other words. And not just them, they were giving her a way out too. We deceive ourselves that we might deceive others better. So hearing me say things would never be the same again, which in truth I said almost automatically to almost every family, however specific or particular it may sound – as a way of changing expectations, but also of buying myself more time – was also something of a relief; the first time she could exhale with catastrophic certainty.

To inhale what, though?

★

As someone who had only just met him I could see what a remarkable recovery Michael had made from a collision that should have meant fatality or, if he was exceptionally lucky, permanent dependency. Apart from the small area of deep pink scar tissue over which his otherwise long black hair would not fully grow, he looked lithe and vibrant, like the framed premorbid photos of him on the piano, on the mantelpiece, the dining table, all along the hall; skiing, diving, mountain-biking in the hills around Bruern, drinking tea in Darjeeling, hugging an alligator in the Everglades, playing doubles on the Heath. I wondered if the house had always been a shrine to this estimable action man, or was it a memorial now, or again, a target to incentivise his return. To the naked eye he looked the same, nearly; only Photoshopped by a lightly malicious teenage daughter, so that his eyes were ever so slightly off centre, his smile a little too wide, the lightest divot – a feathered sand wedge – in the corner of his skull which the cranioplasty had failed to smooth. He shook my hand like he was landing a marlin, asked me if I'd like a drink in a voice that was unnecessarily loud, laughed in the wrong places. Otherwise he burned with health.

And it wasn't just the looks. I'd read the recent neuropsychological screen that morning. His intellectual ability, his memory, his language skills, and most of his executive functioning were in the Very Superior range, above the ninety-ninth percentile. In other words, if you lined up one hundred men, a thousand even, who were the same age – most of them balding, overweight, with back problems, engorged prostates, five decades' worth of accrued worries and regrets – Michael would be the man on the extreme right, the handsome one, floppy hair, waving, carefree, 'Very Superior – over here!', talking as though through a hailer, laughing at his own jokes because they *are* funny – you can't miss him. There had been a reduced performance on a speed-of-information-processing task, some qualitative evidence of impulsivity, and one low score on a planning and organising task. Otherwise all was well, better than well.

While Michael was trying to show me a plan for the croquet lawn and an architect's drawing of the ski slope, Catherine was trying to show me the brain-training software she'd bought for his tablet and the prosaic-looking rehab schedule the multidisciplinary team had designed. (Michael's instinct to avoid brain-training was sound enough; thousands of children's games avowing the stimulation of neuronal plasticity with as much robust evidence as there is for playing croquet on a synthetic piste.) Behind Catherine's back, I would learn, he was using the tablet to swoop for no-fail Bitcoin, or to watch films about big game, then drunk-Skyping his Norwegian colleagues in Guatemala to invite them to his Christmas summer party. I could hear him from the next room:

'These therapists types, they're all *consequences* and *modifications* . . . So fucking de-pressing! Well here's the nugget; this bump on the head has given me a new lease of life, like a control-alt-delete!'

So in appearance, according to the tests and by self-report Michael was near perfect, on top form, rejuvenated. But as he falls asleep next door – profound fatigue a crippling symptom in this patient group, especially those who like a sharpener at lunchtime – the rest of the family will tell me, with increasing frankness, how devastated they are by perceived change and loss. This model husband/father has become 'embarrassing', 'weird', 'lazy', 'foul-mouthed', 'a retard', 'absent-minded', 'clingy', 'a retarded weirdo', 'downright dangerous', 'a full-power semi-automatic alcoholic fuckwit'.

I'm left wondering where the real fault line lies. It can take a long time to establish, especially when the world of the patient is exotic to the clinician. The neuro-cognitive assessment may well not be sensitive enough to pick up on the consequences of Michael's obvious frontal injury, especially given his turbocharged premorbid abilities, especially as he moves around the real world, far removed from the paper-and-pencil tests of the clinic room. And experience with high-end extreme-sports trauma has taught me not to assume grandiosity, impulsiveness and silly clothes are injury-related; they can often be part of the uniform for successful financiers. Equally, family members may have long-extant prejudices which use the trauma to surface. The 'degree of impairment' would often become adversarial, colonial even: both sides finding various pretexts to contest each other's province, and when successful, extending the border a little further into opposition territory. I would have to be patient and find out for myself.

'Whatever you can do,' said Catherine on my way out. 'One thing, don't be fooled by his bonhomie: he's super-ruthless. He fired one guy the week before his wedding. Michael was to be best man. And he fucking loathes doctors.'

'I'm not a doctor.'

'You tell him that, Dr Benjamin.'

I need a quick win to engage him, one that might usefully illustrate the contours of any compromises to self-awareness, impulse control and social cognition; a real test, not some useless workbook exercise which are too often the preserve of rehab specialists. Michael phones me with one: an opportunity, he tells me, too good to miss. Turns out we are both Shaymen – Halifax Town fans from childhood, which has meant an eternity of ever-blackening darkness ... I don't tell him that I long since jumped ship and regularly go to watch trophy-rich Arsenal. The shame of assumed silverware: like winning the London Marathon on behalf of an equine charity in a pantomime-horse outfit, with a real horse inside.

'So, Doc, mighty Halifax are due to play the Gunners at the Emirates in the third round of the FA Cup and we are going.'

'... OK ... Do you foresee any difficulties, Michael?'

'No, not a problem, we should have enough to hand them their backsides.'

'What about for us?'

'What do you mean?'

'Well, the logistics of going to the game, negotiating crowds, managing intense feelings.'

'No. It's a football match,' he says dismissively, as though I have suggested hand-holding in group therapy.

I don't voice the problems I see either, maybe trying to ingratiate myself too much after Catherine's warning. But I do see problems: blood, tear gas, faceless thugs with huge bared stomachs punching police horses, Michael their unlikely capo ...

We meet at King's Cross. My patient is wearing furiously red cycling tights under a Barbour jacket and a deerstalker: he looks like a posh Teletubby on a hunt for snow leopards. We get on the Tube with a group of Town fans. There are thousands of them. I'd forgotten how much they ('we', I mean) hate southerners, cockneys in particular, not that there are any cockneys at the Emirates. They're shouting, they're drunk, they have swallows inked on their necks, earrings in both ears. Michael starts their song for them at the top of his voice:

'I am a Shaymen fan! I am a Yorkshireman!'

... and then they join him:

'We know what we want! And we know how to get it!'

Though very drunk the fans still notice this strangely dressed strange man at the centre of their huddle, arms around them, singing though unsure of the words, shouting in fact, more loudly, more animatedly than they, even without a drink. Notice and then embrace him: Michael, instantly the fanatical heart of Halifax Town Travelling Supporters Club. Just like in my vision.

He turns to me:

'Such good guys, real double-yolkers!'

We get off at Holloway Road. I'm on red alert in case I see any of my Arsenal friends. There are several middle-aged men in Burberry loitering at the station whispering 'Tickets … tickets … Anyone need tickets?' Michael approaches one. He wants £300 for a pair. Michael effortlessly brings the tout down to £200, £170, £150 … We have our tickets. He looks at me: no problem.

'Think this is going to be our day.'

It is not until we get to the turnstiles that Michael realises we are not in the Halifax end, but the North Bank, as far from the away supporters as it's possible to be. He's momentarily outraged, but distractible like a child, and we are all children filing up the stairs into the stadium: the fraction of a second, never quite habituated, when the green dazzles ecstatically, above it the giant field of enclosed air demanding our breath as entry fee. Thank God Halifax will get beat, I think. Their form is atrocious, whereas Arsenal are consistently the fourth best team in the country (except when they're sixth), haven't lost a home tie to lower-league opposition in decades.

More often than not I've been right to shun the risk-assessment ethos that dominates the NHS and my profession in particular. This was not one of those times. Eating our pre-match meal, a £6 meat pie with no filling, I give Michael an avuncular speech about acceptable behaviour, throughout which he looks at me glassy-eyed, nodding in the wrong places, his mind anywhere but here. My own little theatre of the absurd: trying to tell a professional swindler/football hooligan with a brain injury about acceptable behaviour.

'Understood,' his stock phrase for professionals.

But he'd have gone to the game without me, I rationalise, and hopefully I can limit any damage if things go awry. Such is the Freudian architecture of the left temporal lobe that it was only while I was explaining to Michael, moments before kick-off, how we were in the middle of people who were potentially very aggressive and therefore needed to be mindful, that I remembered my professional liability insurance had recently lapsed.

'Mum's the word,' he says, not hearing me.

The referee blows for the start of the game.

'Kill them, Shaymen,' shouts Michael.

He will of course be hyper-litigious in the event of an incident.

Though most of the Emirates is an upper-middle-class morgue, the uncharacteristically noisy little enclave where we are seated seems to consist exclusively of 1970s hardmen in horrible leather coats, who look as though they could headbutt a stone lion to death (Mangan 2018). Luckily none of them know me. But that means we stick out, or rather Michael sticks out; instantly alien and therefore vulnerable. I'm bearing the awareness of this for both myself and my scarlet-stockinged patient.

The whistle goes. I grab his knee, hoping to transmit a sense of self-containment and calm. Instead my hand trembles uncontrollably. Now they're looking at us – my hand on his knee – through their small, pink, homophobic, Halifax-hating eyes.

Arsenal dominate. The crowd settles. I relax a fraction. Halifax score. Michael jumps up.

'Get the fuck in!'

He throws his deerstalker in the air, as though it's Henley, his accent instantly migrated from West Hampstead to West Yorkshire.

'Told you it was our day, Doc.'

So I've established he lacks awareness and has impulse-control difficulties. But then he's a fan, it's part of the job. I pull him back into his seat before the stewards see him. They haven't but it's still too late: everyone else has. A deep female voice whispers into my ear, 'You're dead.'

I spend the next seventy-five minutes on different, equally futile, inter-related projects: persuading Michael not to emote in any way ('I don't know why you're so anxious, we can definitely take them if it kicks off'); persuading the deep-voiced lady and other potential assailants that I am a doctor and that Michael is a patient with an injured brain – I have no stethoscope, no surgeon's mask, no copy of the *Lancet*, and, clinchingly, I'm not a doctor: 'His brain's smashed, like Gazza, proper fucked.'

I don't sound like a doctor either; I change tack, trying to communicate to her and her friends that I'm one of them, mouthing, 'Gooner … season-ticket-holder,' behind Michael's back, except he senses my efforts; 'Just ignore them,' says Michael, putting an arm around me; and willing Arsenal to score.

In the eighty-sixth minute Giroud goes down, a little too easily: penalty. It's right in front of us. Everyone screams. I stop Michael from remonstrating with the ref. Arsenal equalise. Four minutes left plus stoppage. I think it best to leave. I usher my patient to the steps, head bowed and lightly shaking, as though relieved a disastrous loss has been averted, as if to say, 'No point enduring any more.' Nobody's buying it. They glare, waiting for one more false move, one lapse in the performance, one more hiatus of impulse-control. Someone throws the remains of a coffee (£4.50) in our direction. 'Flat white: typical Arsenal,' I think. Michael doesn't notice. Somehow we're allowed to leave the stadium.

'We'll nail the southern twats back at the Shay,' he tells me the moment we're outside. 'You've got to get tickets, Doc.'

There is the lack of awareness caused by damage to self-reflecting modes of cognition. Then there is denial in the face of emotional trauma. They often go hand in hand in TBI. It's no surprise that Michael won't accept the football match as evidence of problems. Restoring awareness, when its successful, isn't a light flicked on, but the slowest of dimmers. In the following session I set him another task. He wants to book flights to Valparaíso to visit a former colleague; the kite-surfing is 'out of this world'. Too dangerous I say. 'Chess, then. In speedboats … Come on, it's only an exercise.'

He researches the cost of a first-class ticket to Chile on 12 December, returning on 16 January. Later that afternoon I get a call from Catherine, obviously distressed.

'I'll tell you what's wrong Dr Benjamin: he's just bought thirty fucking thousand pounds' worth of airplane tickets to Jakarta, leaving tomorrow.'

That same evening the three of us are sitting round the kitchen table.

'He just can't be trusted.' She's looking at me.

'We needed a break.' He's looking at me. 'It's been a difficult few months for all of us, and it looked so … mesmerising,' he explains.

'Mesmerising like the church organ you bought? Or the Russian hunting dogs? Or the fake Rothko?'

'Investments, sweetness.'

Lack of awareness of impulsivity or denial under attack?

Such is our genius for self-deception, there is no behaviour under the sun that cannot be denied or distorted. Most doctors deny this of course, believing they can corner anyone into admission with the force of their reason. But the cornering is the problem. With Catherine gone and a gentler style of enquiry, a down-at-heart Michael explains how sitting at the computer, 'totally clear' about his task, a brightly coloured box popped up announcing an *urgent* airline sale: *'Heavenly Jakarta, jewel of the East: The mesmerising 7 star Kapoor … The private, tantalisingly white sand of the Indian Ocean footsteps away … Only six tickets left … Only five tickets left …'*

One click and the computer's memory of Michael's previous bookings do the rest. Technically it's known as stimulus-bound behaviour; the damaged front lobe means the person can't unstick themselves from an alluring trigger – like four-year-olds and fizzy sweets. The interface with global marketing syndicates expedited by idiosyncratically supportive computing is not the perfect environment to retrain these habits. Fortunately I was able to get the airline to return the money.

Later Michael confesses in private, 'I couldn't resist ... I tried to, really hard, and I just *couldn't* ... I don't know what's happened to me.'

It turns out that Michael's awareness – his readiness to acknowledge a difficulty as being his own – depends to an extent on interpersonal dynamics. Some of this is obvious: if there's a high social cost he is less ready to admit something. If my tone however becomes jocular, laid-back, less expert, then Michael, who doesn't like doctors or authority in any form, is more ready to explore the possibilities that things are no longer the same. There is an inversion of the effect seen in our research with the medical students: the less of a doctor I am, the more of a patient he can become.

Slowly we develop a way of separating Michael from his brain injury ('the fucked bit'), allowing it to take responsibility for certain difficulties when framed lightly. In this way a small, oblique but secure-seeming bridge is built from Michael to an otherwise devastating self-awareness.

Sadly the family, in tremendous turmoil of their own as they reconfigure their identities around an absence, haven't got the time or space for 'subtle', 'nuanced' approaches, and are compromised, like the rest of us, by the extent of their own self-awareness. Instead, every mistake, like the scores of framed action photographs that decorate the home, reminds them of the stature of their premorbid father/husband, of all they have lost. If they could only put all the photos together in the right order in another scrapbook and flick through it quickly enough, the old Michael would jump back to life. As it is, the 'fucked bit' is all they see.

Over time a further subtlety emerged relating to denial, which, was all but impossible for me to see even though it's the most obvious thing imaginable. At its most naive and linear there are two fixed points, two photographs: 'before' and 'after', 'pre' and 'post' injury. The work of rehabilitation is to reverse time's arrow, bring one as close as possible to the other. But to do so requires fixing the meaning of both. As I got to know Michael and his family, I came to understand that just as there were competing versions of what he was like now, there were discrepancies in what he was like before, which are only permitted to emerge as the magic dust of 'two years' loses its sparkle. Like my patient, I had been in denial, had been totally seduced by a fantasy of a pre-injury Michael, a flamboyant, irrepressible man who would accompany me to football matches (the antipodes to my introverted, depressive, football-indifferent father – preserving him, preserving this difference, was therefore in my interests too), the man who could stop time, who had never done anything out of character, was always just 'Michael', through and through. If 'after' won't move any further towards 'before', if fifty-eight can't be fifty-five, let alone twenty-two, then it may be that 'before' has to move closer to 'after', to turn back to the future.

A different story was slowly allowed to surface: of someone who had long been sick of his job, bored of his colleagues, had pined for something different, who wasn't so Michael after all. Turns out this Michael had lived in tacit expectation of a catastrophe for years, dominating his world, making oodles of money (more stimulus-bound behaviour but without the brain injury) as a means of ignoring the panic attacks, shouting his way through social anxiety, constantly in motion to keep the depression at bay. In other words, his 'youthfulness' – good genes apart – his high spirits, his character, his *juice*, had become an affectation, a defence against desperation. And the family were complicit, enshrining the Michael they want and need. Then a real catastrophe arrives in the form of a head injury, and instantly his acquired symptoms, though to the naked eye a close bedfellow of what went before – brashness, thoughtless extravagance, opinionated unconcern – are now rigid, unaffected, imposed. It was as though he'd become his own not-quite-synced puppet or foreign film, separated by an uncanny valley from his former self. And, cruelly, the qualities that had once been so seductive become the very things that estrange him from others. Or that's a story, formal enough in its irony, for clinicians to tell themselves and feel satisfied with. But really he was long estranged from others, and crucially from himself, by the same or similar attributes, which though *looking* natural were more or less performed. Viewed as such, the line between 'psychological' and 'neurological' starts to disintegrate – they depend on the same mind after all.

Most weeks I see people, helmeted or otherwise, die from head injuries like Michael's. Then there are many more who will spend the rest of their days with the most profound physical and cognitive disabilities, rescued by the latest surgery to lives of dependence. Michael got everything back, nearly. Over two years he was told that he's a miracle by a cathedral of medics and therapists. But gratitude doesn't last; or it's not enough. The shortfall, the denial and lack of awareness of it in him, and the denial and hyper-awareness in the rest of the family, meant that things would fall apart. The Literature suggests that up to 50% of couples separate in the first two years of a TBI, partners routinely describing personality change and unawareness, rather than physical or cognitive devastation, as among the main reasons for their separation.

For Michael there was a further hint of something that could never be spoken about, certainly not photographed. The injury, terrible though it was, removed a load, offered him a way out of a life that had run aground, instantly recast him in a different idiom, against 'health' perhaps, but more authentically; returned him to a different future, still like Michael J. Fox, but the older Parkinsonian version. 'Like a control-alt-delete,' I remembered his description. The head injury created new limits that meant he was incapable of paddling hard even if he'd wanted to, took away the possibility of looking effortless, keeping up, staying young. It wasn't intentional of course, but the trauma was literary in its efficacy, giving him a darkly elegant permission to change track.

But still that's just a story, too neat in its telling, ignores the mess of suffering it can't contain.

A few months ago, in spring 2018, I met Catherine for lunch. It had been years since I saw her. Though not typical for my profession, I would on occasion spend time with certain patients outside of formal clinics. In the case of the Taylors I had worked closely with them for many months and then they disappeared. I missed them, Michael in particular, was curious to find out how they were doing.

She'd managed to 'stick it out' for the 'two' years, when all hope of further recovery had elapsed. Michael was not coming back any more than he had, and that wasn't enough, or it was too much. She'd found a peace of her own in a humbler station, away from the painfully strained routines of an executive's wife. Never mind being the same again; Family Taylor no longer existed.

She placed another photo of him on the lunch table, for my scrapbook – next to the graduation and the fifty-fifth birthday – taken the previous month by one of the children at a marina near Wapping. Michael, wearing a panama and filling a large salmon-coloured linen suit, was trying to board an amphibious Cessna which would drop him outside his new seafront home in Bournemouth. Behind him, a porter had dropped his luggage so that he could give Michael a push up into the plane. Michael was corpulent, unsteady on his feet, suddenly in advanced middle age (suddenly *like* my father), about as far as it was possible to get from the 'pointing' photos of most of his adult life. No change for thirty-plus years and this powerful sense that the world would yield however he chose to test it or himself; and then this – in less than five years. He was laughing his head off at the ridiculousness of it all, I could see now.

But only because his 'all' was smaller or at least different from ours these days. Catherine told me he loved sitting by the sea. He'd lost the taste for adventure. He drove his Maserati to the pub for lunch, had a maid, fished occasionally but preferred watching, was dating a high-end estate agent (she'd sold him the flat). The kids accompanied him on summer holidays to Jersey, spent a few days with him at Christmas, and called him when they needed a loan.

Jane

A series of friezes like the Stations of the Cross: left arm stiffening, jaw jutting – speech marooned mid-sentence – eyes disappearing upwards, trunk listing. They became more frequent with the progression of puberty. The timing of it couldn't have been worse; the 'look at me/don't look at me' of that age, the microwave cruelty of young girls fed by, in all probability feeding, her hourly minute-long jigs. Six months in and she felt like she'd always been epileptic, that it was a manifestation of who she really was, it had just waited a few years to come out.

Like her, Jane's father found the epilepsy embarrassing. Even though he'd learned not to say it, she could tell he thought she was being 'theatrical', that the displays were somehow indicative of poor self-control, further evidence for his signature quip that she'd been adopted at birth. Jane – scowling, clumsy, raven-haired, big-handed, a natural comedian – an Addams in a family of Kennedys – like satirical quotation marks around the rest of them. Our twice-yearly reviews were photographs in a protracted time lapse. I watched her strobing from a gangly, disarming, unscathed eleven-year-old; clothes blackening, hair changing colour from black to blue to red to black again, the end of laughing (she believed it might trigger a seizure), piercings pricking her nose, top lip, eyebrow; a tattoo of a dolphin, then a fairy, a game of Quidditch – the beginning of a mural, which spread in staccato across the inside of her arm, then the outside, the shoulder, her back; a cyclops, Jesus with breasts, Marilyn Manson lyrics, a harpoon through the dolphin's snout, a noose round the fairy's neck. Meanwhile, the disease has caught up and overtaken the young woman, as though bound by different laws, faster than the years themselves, with the rest of the family disappearing from view, one by one, in sudden abrasive jumps.

As a child her father had told everyone he wanted her to study medicine, 'to look after him' he joked, but he meant it. But she left school before her GCSEs to work with children with learning disabilities as a basic assistant. Many of the children had huge physical and mental challenges, and often the most spectacular seizures. Being with them made her feel like she had something to give, and could, on a good day, gave her a feeling of relative control. At fifteen she started going out with a twenty-three-year-old who was nothing like her father, the start of a pattern that would play out over the next few years. At seventeen she left home to live with one of them. She learned how to drink and take recreational drugs without throwing up. She believed, or said she believed, that mixed in the right way they took the edge off her seizures. Unlike the anti-epileptic medication; she'd taken a dozen medications at one time or another, at high dosages, with as many as four different ones at the same time. There was always nausea, fuzziness, additional memory loss (her epilepsy caused amnesia), and sometimes more specific side effects – hair loss, hair gain, weight loss, weight gain, dry itching skin, voice-deepening, pain in her limbs, disrupted menstruation. Even if they didn't really stop the epilepsy, the booze and weed didn't do anything like that, the only obvious side effect was the paranoia about fitting when she was smashed.

It's twelve years since she was first diagnosed, estranged from her family, working fourteen-hour shifts on minimum wage, single again, the seizures identical to how they've always been, if not worse. These days the kindly epileptologist's tack has hardened, appealing directly to the adult in her: 'Imagine a world where things are worse than they are now, more seizures, worse seizures, the world without medication.' But Jane finds it hard to use her mind in this way, to judge herself against other, virtual selves, and then be grateful for it. It may be that her capacity for imaginative abstraction (left pre-frontal areas) has been knocked off by years of seizure activity; or that her capacity to generate feelings of self-care (right pre-frontal areas) has been similarly compromised. More likely she just hasn't got the space for feeling grateful when after half a lifetime on the disease's brutal leash she is just sick to fucking death of fitting, prescription drugs, slurred speech, mental vagueness, feeling fat, social anxiety, cannabis dependence, being fat, binge drinking, transient hostile codependent relationships, horrible crap tattoos, early dementia brought on – she is convinced – by the 100,000-minute-long assaults on her brain. Having first raised the possibility years ago, she is now imploring us to consider her for the epilepsy surgery programme.

Epilepsy surgery is the only instance in medicine where a person under no immediate threat volunteers to have a brain injury. To offer a surgical intervention, the experts have to weigh up the evidence to the point where they are convinced that whatever the potential implications, surgery represents the best option for reducing seizure activity. Without this there can be no *informed* consent. And 'informed' here should be caveated; we are only beginning to gather the data required to make predictions about the 'possible' (i.e. inevitable) neurological consequences of removing portions of the brain. There have been mistakes: we once turned a pre-surgical chemical engineer into a post-surgical petrol-pump attendant who still suffered terrible seizures. Which is to say medically deliberate brain injuries are in their infancy; wisdom takes time.

One might also ask if someone like Jane, who has been so credulously wedded to the idea of surgery for nearly a decade, is actually *giving* her consent rather than thoughtlessly demonstrating a reflex. Can she weigh up potential consequences, especially, as mentioned, when this weighing up may itself depend on the exact same neuroanatomical areas that have been compromised by the seizures? Injury apart, could any of us really imagine what it might feel like to have fundamental aspects of who we are erased, without our imaginations turning us into someone else. Try it . . .

For clinicians and patient alike the decision is a partially sighted *pas de deux*, if not dancing in the dark. It's the clinicians who lead, of course: before Jane gets to make her decision, before she finds out if there is a decision for her to make, we make ours. 'We' in this case means a minimum of two epileptologists, a neurosurgeon, a neuropsychologist, an electrophysiologist, a neuroradiologist, a clinical fellow in epilepsy research, an epilepsy nurse specialist and a neuropsychiatrist. Together we address questions such as whether or not the seizures have an epileptogenic focus (coming from one specific part of the brain). Or are they distributed across several areas? Will her brain be able to compensate for any cognitive losses? Will her language be obliterated? Is she sufficiently resilient to cope with the consequences? These are questions which require multiple investigations, often taking months, occasionally years, before they are brought to the surgical meeting.

★

A darkened room, as many as twenty specialists gathered round a screen as big as a cinema – consultants on the front row – onto which are projected giant, shimmering MRIs. There are no film crews allowed in here, no patients either, only images six feet tall of the insides of their disembodied heads:

'It's a grade-four glioblastoma on the sagittal view of the right maxillary body … Isn't it?'

'Isn't that a sclerotic posterior fossa?'

'I think that's Ken's hair catching the projector.'

First one to shout out the right answer, like the picture round in a pub quiz.

The drive towards medical specialisation has meant we can only make certain decisions by combining expertise. We assume consensus can be routinely achieved without acknowledging certain threats; intolerance of uncertainty, confirmation biases, inter-professional power imbalances, interpersonal differences specific to the group. A superficial example – superficial because God only knows what goes through the man's mind – at one meeting I used to attend at a hospital in the Midlands. I could set my watch by the fluctuating blood sugar levels of the massive surgeon as they spiked and then crashed following his ingestion of a large Danish. Meaning that after initially splurging his opinions everywhere, he was unconscious during the more prosaic, detailed discussion of the current evidence base for the chemical treatment of, say, a paediatric brain cancer. That discussion is led by the oncologist, a human PubMed (the doctor's IMDb) of relevant and recent research which he rehearses in unedited, uninflected, monotonous wealth, at a frequency and tone that is powerfully soporific. When the conversation eventually wheels back round to the sleeping giant and the room canvasses his perspective on a possible surgical resection in the light of this evidence, there is only silence … The long-suffering oncology nurse specialist, sitting next to him throughout, the only person aware of his mini-coma, had, I imagined, considered rousing him with her red biro, but because, I also imagine, she longs for a retributive humiliation after years of his maligning or worse, ignoring, didn't ... So silence, until a myoclonic jerk, itself a proto-seizure common to the well, and influenced by, I picture, a surgical-cum-pastry-inspired dreamscape, rouses and seems to produce from him a noise which sounds like no known word, 'Flefffflleefffff ...'

Which is then clarified, under the weight of gathered expectation, with typical face-saving brio: 'Fffetch him in ... Her?' says the surgeon.

And the decision is made, the knife is readied, and nobody who was there can quite say what just happened.

So how can the status of this particular neurosurgical decision-making algorithm be assessed? The design of that audit would require some kind of multidisciplinary super-convention: Roman Jakobson for the discourse analysis, Wilfred Bion for the group dynamics, Ilya Prigogine for the modelling of chaos, Venerable Chokyi Nyima Rinpoche for the exchange rates of interpersonal karma, David Foster Wallace for parsing the recursive contribution of irony, etc. (As things stand the Trust relies on honorary – unpaid – psychology students.) Even then, could these tonal frequencies ever be heard, let alone audited, these shadows ever be truly *seen*, when the corridors are stalked by TV crews and the invisible men and women from the Care Quality Commission; when the Trust ideals – Excellence, Kindness, Dignity – loom above us from giant Riefenstahl-style posters, insinuate our prose as tiny slogans programmed to automatically appear at the bottom of emails, letting us know that what we are seeing is ourselves being watched?

Something more primitive still threatens the most basic assumption: that our experts can discuss issues in an 'adult', 'professional' manner (or whatever sanitising metaphor you choose), that our secret life will remain undetonated. We need to believe that character and autobiography will not contaminate medical decision-making, that we can partial out all that is rogue, split our personalitites in two in the name of professionalism. Which is by way of a confession. In Jane's case, I was not without prejudice. I have a brother who has suffered with epilepsy since the age of sixteen. Like Jane, there couldn't have been a more difficult age socially, only worse than Jane in severity, my brother suffers generalised tonic clonic seizures, meaning a total loss of control. He first announced his epilepsy in full school assembly with a two-minute seizure. Other notable fits have been in a club on indie night where it was mistaken for an Ian Curtis tribute; in the theatre during a Pinter Pause; and while driving a car at 40 mph – one seizure means two years without a licence, for good reason. The consequences for him were complex, but shame and loss of independence were prominent.

It affected me differently. One afternoon three years after his first seizure I was having tea with a friend when I was suddenly struck by the sense that my words were no longer making it out into the world, co-ordinated with the sight of the friend cocking his head like a dog at a novel sound. Which is to say I was aware of seizing, its bending of reality, like there was something still deeper inside watching the thing inside, which is a giveaway diagnostically. The situation was an extreme one; final examinations were days away. I, along with thousands of others, were at a limit of anxiety and exhaustion. But I assumed, as it came after my brother's diagnosis, that this occurrence was the familial expression of epilepsy. Being simultaneously hypochondriacal and a fundamentalist illness-denier – like most young men – I never got it investigated.

I now know it wasn't epilepsy but something else. Non-epileptic attack disorder has a prevalence rate of 2–33 per 100,000 which means it is not uncommon. People have seizures which to the lay eye look like epileptic fits and as far as the sufferer is concerned they are involuntary. Only there is no epileptogenic activity on EEG. Neurologists will point out subtle variations that distinguish them from the semiology of bona fide epileptic seizures: duration, subtle retentions of physical control, eye movements, pockets of awareness, etc. To complicate matters further, many people who have epilepsy also suffer from non-epileptic attacks, the latter piggybacking on the former, learning from it, ventriloquising it, so that distinguishing between the two may be impossible for the patient. This is one reason why it is notoriously difficult to treat – the treatment here is psycho- rather than pharmaco-logical. Another reason for the difficulty, at least in a significant minority, is that the patients have undergone severe trauma in early childhood which has, in some remote, glacial way, made them susceptible.

'That's theta spiking.'

'She looks depressed.'

'In the left frontal gyrus.'

'No sugar, please. Can we have the three-tesla contrast again?'

'Tell her it's alpha not theta that's doing the spiking.'

The darkness allows us to be more ourselves.

'How far is it to the Cotswolds from your place?'

'Tell her yourself.'

'There's nothing frontal on the 3T.'

'You tell her.'

'Is there?'

'I can hear you, I am told.'

'Who are we talking about?'

'Depending on A40 contraflow.'

The various lines of investigation create cacophony in the twenty minutes allocated for her case. I wonder if people are actually hearing one another; I wonder what is really on their minds. I think about Jane. I think about my brother. I think about those times in my life when I have been a patient, and how vigilant I am for any sign that the person I'm talking to might not be listening. As the conversation weaves in and out of the different findings, a narrative evolves subtly at first, then with more solidity: there may be a non-epileptic element to some of her seizures, slowly growing up alongside the epilepsy itself: like an alter ego, a purely 'theatrical' her (the one recognised by her dad) which might be the unconscious expression of not being heard.

Why won't we hear her? It's not that we haven't got the data; we have all the data in the world these days, we just don't know how to constrain it. Diagnostic reasoning, even when people are listening and in control of themselves, is prone to bias. Some studies suggest that clinicians don't reason at all, but base their decisions on 'clinical experience' alone: no meticulous hypothetico-deduction here, instead a base, unsophisticated recognition of previously seen signs and symptoms. And if there is reasoning, the tendency may not be to rely on nuanced logic but to fall back on obscenely simple rules or heuristics which are often deeply idiosyncratic and therefore distorting. Really we are precocious children trying to play Snap in a hall of mirrors.

I am doing my best to listen, to hear her, which means fighting off a whole associative complex: the fixatory pull of the lone half-inch hair sprouting from the nose of the neurosurgeon; the nauseating print on the epileptologist's two-tone dress which from a distance appears to flash the word 'ham' over and over; the stilted conversation I had the last time I spoke to my brother; sadness at what I might have been communicating all those years ago, un-hearable to others, unheeded by me, while we discuss inconsistencies in the data, speculating if these extra pseudo-seizures are Jane's coercing of us to let her have surgery (doctors hate their patients' coercion more than anything else; usually it's they who do the coercing, making sickness a form of resistance), as if her voice, notable by its absence today, were not enough on its own. If we are really here, then it's our job to listen.

There are always two experts in any given consultation: the patient and the doctor, one skilled in the particular experience of symptoms, the other in investigating them, first- and third-person accounts vying for the same conceptual ground. The competition takes place behind the facade of trust – that the aims of the doctor and the patient will ultimately coincide. But the outcome of a consultation often reflects the strategies used to establish authority rather than clinical need. No holds are barred, the patient resorting to a huge unconscious symptom-display; a rebellion expressed in Charades, because she has no words left with which to petition us.

'Why are none of the nurses speaking to her on the telemetry?'

'Looks like a panic attack.'

'They are Agency.'

'Milk?'

'And the margins of the parieto-temporal cleft?'

'Ouch.'

'I don't believe her semiology.'

Jane, you are not here but I can hear you.

'Quiet, please,' I say, 'just for a moment ...'

There is silence.

As with many instances of unconscious communication, the effects of the pseudo-seizures are paradoxical: they tell us, who are untrained in, largely deaf to, the real meaning of such messages, not to recommend her for surgery, to refer her instead for some specialist treatment in trauma-focused psychotherapy. She's likely to be more distressed by this, feel like her disease which is too much for her family is also somehow not enough. So that now we are another family who won't accept her for what she is, won't 'adopt' her, because in our systematic, considered and expert way, we do not believe her.

'Let's just give her what she wants … Come on, just this once … Take out the diseased bit, it's only the size of a finger-joint, she won't miss it *that* much … Guys … ? Please?' is what I want to say.

Rooms

There is always a room with two people.

The rooms change. This one is a small octagon made of native redwood. There is a bed in the corner, an altar, a shelf with a few books – a psalter, Meister Eckhart, *The Brothers Karamazov* – a miniature Bose docking station for a smartphone, otherwise it is empty. We face one another, cross-legged on zafus.

There is always a room with two people, one of them is talking.

'It's like I have a stadium-filling drum kit inside that belongs to someone else. It's rolling the moment I wake at 2 a.m., getting juicier the longer I'm up. I have to keep my mouth shut tight, hold my breath, otherwise ... sshhhhh ... I could never whisper at school.'

He can't whisper.

'I would get confused between thinking and thinking out loud. Is there a name for that?'

'Er ...'

'It has meant a lot of shit. The teachers heard everything, heard stuff *about* them. My mother's the same. Is it genetic?'

'Er ...'

'Why is it that monks are less likely to get Alzheimer's?'

'I don't think we have a full explanation of ...' I'm about to reply but there is no time: he doesn't need to breathe, apparently. On him the shaved hair has a skinhead feel.

'By seven-thirty it's near the top of my throat, like this mad animal is thrashing around ... That's five hours keeping it in, all the way through vigils, lauds ... The psalmist says, "I go to sleep crying and I wake up laughing," like *he* had bipolar, like we each go through every different diagnosis in the course of a single day ... The moment I'm out into the Big Sur air and open my mouth and breathe ... BAMABAMDAND-SMASHSNHSIANASFKFNCHACHACHAA ... Do you know what I mean? BASHASHSSHSHSAASSHSWHHACHA CHACHAA!'

The walls absorb the sound. I do know what he means: he is in ecstasy. Sometimes we envy our patients.

The hermitage is a pop-up, erected last year to replace the disintegrating 1950s breeze blocks – the Order, Italian originally, is over a thousand years old – on a postcard plateau a mile above the Pacific. As well as state-of-the-art soundproofing, the outside of the redwood consists entirely of solar panels: Californian monks.

'I need music. It has to be something super-quick, not thrash, but punkish, like track eight on the Strokes' first album. You know it? If I can make it down the hill to Pacific Highway One – which is exactly two miles – at that tempo I know I'm ripping it. And if I'm really feeling it, then it's 'I Wanna Be Your Dog', I did it in just over ten minutes once with that on a loop. I'm belting it out all the way down, BABABABBABBABBBABBBABBABBA ... *So messed up, I want you here ...*'

He is singing.

'That's not everybody's idea of being a monk,' I suggest. It makes no difference.

'I feel the warmth of dawn, pockets of chill from the sea, I smell aniseed, marijuana from the farms hidden back there. I'm flat out, it's so steep everything is lifted up to my face, each stride is gigantic – fifteen, twenty, twenty-five metres long ... Why not? Those fell-runners from Yorkshire do a hundred yards in less than five seconds. It's the music, we call them backing tracks, but without them nothing stands out. It's our hack on being alive, it tunes who we are; it changes everything, e-ver-y-thing.' He pauses for a moment.

It changes everything I thought. The room changes everything too. Here, in *this* room, he fills the space: my questions can't be asked. Is that his intention?

At different times in my career I have worked overseas; India mainly, but also Nepal, and North and Central America. I spent two years in California, a few months in LA with Latino gang-bangers, with transgendered migrant workers in San Francisco's Mission, with crystal-meth heads in Fresno and the homeless in senescent super-rich Carmel. 'First World problems' you might call them.

In Big Sur I worked with Bede. He had a complex psychological history that we only skirted around. Two years previously, then aged thirty, he was suddenly, impulsively inspired to join an ancient monastic Order, without knowing much about its beliefs or demands ('It depends what you mean by "Son" and "of" and "God"'), taking the name as a rite in his newly elected identity. Now he was deciding whether or not to make a lifelong commitment or leave to do something different. The prior explained that for all his quiddity and restlessness his fellow monks held him in great affection. I was to chat with him rather than engage in anything formally clinical. It was hoped that with me being agnostic and secular, Bede, who tended towards iconoclasm and scepticism when surrounded by his brothers, might be more free to feel out the contours of whatever faith he had. Less explicitly, it was hoped I would be a part of a process by which he chose to make Solemn Profession – a monastic marriage – to the community. Numbers were running low and heading in only one direction.

'I'm lighter than anything. I can hear elephant seals head-butting one another on Sand Dollar beach ten miles away, the *whoosh* of the brown pelicans way below at Limekiln, grey whales groaning fifty metres off the shore at Kirk Creek; I am IMAX, high-definition Dolby, whatever.'

Grandiose, deluded, for sure. But if our attention to the ordinary around was sufficiently interested who knows the limit of our sensory-perceptual capacity? We might hear the grass grow or the squirrel's heartbeat (after Eliot 1871).

'The animal is out. I've left the road, jumping high over the pampas and cactus, my feet making deep imprints in the rock, somehow finding ground by themselves, the perfect place each time, to the rhythm of the song. I'm not thinking any more, not a single thought in my head, except my body is thinking and the calculations are stupendous. I feel stu-pen-dous. Our bodies know how to live, our brains make us stupid ... Have you got kids?'

'Um, no ... I'd like to one day, but ...' is all I can get out before,

'There's me and rabbits, lynxes, snakes, pumas and out there [indicating the sea] seals and whales and great whites: all procreating, one giant fuckathon, one big blue fucking planet: we are all at it, constantly, except the monks, because we *are* just creatures, Iggy wants to be a dog when he is already a dog, we are all just dogs ... except for cats.'

Thought disordered but wasn't this just the flood of desire? And what is spirituality other than one response to a surfeit of desire?

I noticed a text carved above the redwood door.

Sit in your cell as in paradise. Put the
whole world behind you and forget it.
Watch your thoughts like a good
fisherman watching for fish.

Faith was present, I thought, but a natural faith. He can't
sit still for a moment, his mind so many flying fish. The cell
spits him out each morning. Paradise is outside, the view so
compelling that the monks had to put boulders on every
switchback to stop people driving straight into the Pacific,
accidentally or otherwise.

'... and I don't know how I'm ever going to get back
again; the sea just keeps coming. I couldn't stop even if I
wanted to; I don't want to. If I keep running in one direction
I'll hit Patagonia or Alaska. I could swim to Japan ... Are you
hungry?'

'...' He's getting faster and faster.

'I dream of fighting a cougar – I reckon I could take one –
or being eaten by a whale, like Jonah, I mean really dream
about it ... Riding inside the thing. The stomach on those
guys has to be as big as my cell right, at least, right? All-you-
can-eat sushi ... Next stop Baja before she pukes me out ...
Don't give me that look.'

'What look is that?' I mean to say, but he answers before
I ask it.

'The risk-assessment look. For the first time in my life ...
When I was seven or eight my father dragged me up Simon's
Seat. We didn't have much in common, neither of us really
wanted to be there. The moment we got to the top, the clouds
parted and all the dales and moors were suddenly there –
bruised greens, bleeding purples, thrilling, all of it – and he
says, "Well?" ... and he waits ... and he waits ... and I didn't
say, "Not bad, Dad, not bad, but not enough," but that's what
I was thinking, that's what I wanted to say. Maybe I did say
it because I thought it ... ? Well, and now for the first time

in my life this – this, *this* – is enough. Sometimes it's too much ... So, what else? What else ... ?'

It is hard to keep listening. It is hard not to listen. There are always two people in the room; one of them can't stop talking. Bede wasn't the first bipolar-ADHD-narcissist-addict – whatever he was – who had gone to town on his Higher Power. But he was the first I met who had gone to a monastery on it. I really did envy his high feeling, his boundlessness; who in their right mind would want to be cured of that?

'... What else?'

Finally he stops. He is breathing hard, looking at me, waiting ...

'What else ... ?' Still looking at me and I am thinking how patients are often our imagined others, that this is how my life could have turned out too.

A different room, in East London where a real-life Jonah lives.

He had been described as 'violent', 'manipulative', 'labile', 'severely impaired', 'paranoid', 'extremely risky'. I was told he spoke no English apart from the few catchphrases and curses he'd picked up from the television. Did this prepare me for the sight of an overweight Haitian man in an aluminium foil suit, an outsized bowler on his head, reflector aviators, multiple gold chains around his bared chest, spurred cowboy boots one of which has been built up, walking with the aid of a black cane crowned with a silver death's head, opening the door of his flat, flinging his arms around me saying, 'Welcome to the A team,' then rejoining three different conversations in three different languages on four mobile phones, one of which was obviously a toy? I wondered if there was anyone on the end of the lines.

'Th ... th ... th ... that's all, folks.' I thought he was finishing one of the phone conversations, but it was meant for me. He closed the door and disappeared.

I looked through the keyhole and shouted his name, 'Jonah? Jonah?'

Nobody answered.

'Jonah?'

'Let's go *through* the keyhole,' he shouted from somewhere inside his lair. Nobody came so I looked through.

In 'The Fall of the House of Usher' it takes most of the story to realise that the eponymous house is the inside of someone's skull. Not so with Jonah's Walthamstow flat – the recognition is instantaneous. There, in the narrow sulcus-like hallway, is a monumental tangle of baby-rockers, bouncers, high-chairs, children scooters – like a Turner Prize-winning sculpture called *The Nanny State*, only I knew Jonah was a middle-aged bachelor. Off to one side, his bedroom. The bed is made up of a half-dozen double mattresses stacked on top of one another, leaving one and a half feet of space between them and the ceiling. Like the bed in 'The Princess and the Pea'. This club-footed princess (I read that he once wore a fairy outfit to a case conference) has an eight-foot stepladder propped against the side to enable her to climb up. On the floor there are half a dozen or more large suitcases packed with outsized clothes (mainly shirts printed with plant-life), aftershaves, soda streams, salon dryers, hair-straighteners (Jonah was bald), as though ready for a world tour. He has not left London since he arrived more than twenty years ago.

Moving down the corridor, on the other side we find the guest bedroom, a room so crammed from ceiling to floor, from wall to wall – with antiquated domestic furniture, industrial hardware, garden equipment, office stationery, fairground cast-offs, an ancient photocopier wedged against the ceiling, a rusted rotavator, a dodgem car climbing a wall – that it looks like an anti-gravity chamber. This would lead to a neuro-psychiatric diagnosis of 'hoarding', an adjunct of obsessive-compulsive disorder, 'explained' to Jonah via an interpreter. He agreed, through the interpreter, explaining back that one day he planned to ship the suitcases, the high chairs, the mattresses, the furniture (the dodgem he planned to make roadworthy at some point) – all tat to the untrained eye – to Dominica, Cuba, Antigua, St Kitts, where he had twenty-three children, mainly daughters, to fifteen different mothers (assuming each of them would by now have a brood of their own – a nursery school of mini-Jonahs when gathered in one place, hence the collection of baby paraphernalia in the hall). They would appreciate these choice luxuries ('Women from my country prefer aftershave'). Jonah's hoarding made sense enough as an expression of clinging to reparations he might make to long-abandoned children. It was more than symbolic, but the cost of shipping was way in excess of the value of the goods, which didn't make sense to him. His insurance case was pending a decision; in his mind he was the virtual CEO of a multimillion-pound business – My Head Injury Ltd. He was on his twelfth case manager in three years. 'You're fired,' he would say, once every three months, pointing his stumpy finger, the only bit of *The Apprentice* he actually understood.

At the end of the corridor we find a small living room, made smaller by three full-size white leather sofas and a desk, meaning it had to be climbed in and out of. There are faux lion, leopard and zebra skin rugs over every inch of floor. He told one occupational therapist that he planned to 're-wild' them on his next home visit. One enormous flat-screen television is always tuned to Al Jazeera in un-subtitled Arabic. Several smaller acolyte TVs are on different music and sports channels. To the observant eye, buried amidst all this, like an imaginatively curated museum for neuro-rehabilitation technologies, are whiteboards, diaries, calendars, card indexes with laminated instructions, palm pilots, Dictaphones, tablets: fateful monuments to the hundreds of hours spent with dozens of different clinicians and therapists who sought to bring order to his world. Nothing had taken hold. Jonah could have bought his own ship to Haiti with the squandered funds.

So, if the house and rooms *were* Jonah's cerebrum, its morphology had never been seen before: a Brodmann area lottery, an exquisitely designed torture chamber for any would-be neuroscientific cartographer.

The discerning eye might also notice the multiple CCTV cameras stationed high in the corner of every room. This was the reason I had been asked by the case manager to join the multidisciplinary team. Jonah had developed the belief that people – former business acquaintances, clinicians, religious leaders (the cast changed every few weeks) – were entering his flat when he was out, or when he was in, sometimes as themselves, sometimes as others, sometimes in non-human guises: spiders, snakes, rabbits; even taking inanimate forms – smoke, water, invisible toxic gas. Once in, they would either remove precious objects – an overhead projector, cheese, a pram – or most often just move them around. The imagined motive: to confuse him. This in a man so radically disoriented by brain injury he can barely tell night from day, whose memory profile is the equivalent of dense Alzheimer's. In response he would reel off cartoonish pidgin curses at the assumed culprits – 'I will boil their heads, peel them, feed their brains to vipers'; 'I will turn their eyes to stone'; 'I will give their children Ebola' – bulging cranial veins, spit geysering from his mouth, obviously enjoying himself: 'I will shoot them on a cold, dreary night in Bucharest' (this one learned by heart from a John le Carré film).

And in a scarcely credible misjudgement one professional had the thought that £50,000 worth of CCTV cameras would help resolve the confusion. As though Jonah was interested in having his reality tested. That his reality was fundamentally immeasurable had been known for many years. He was forty-two. Or sixty-three. Or fifty-one. He was from Haiti, or Cuba or Jamaica. He sustained his head injury five years before, falling or being pushed or jumping out of a slow-moving bus or a fast-moving pushbike. Prior to his injury he was an accountant, a car mechanic, a semi-professional basketball player or a chef, or all of them at the same time. Intellectually he was either well above average or well below it, which still might make him well above a different average – that of an illiterate son of illiterate farmers … And on it went. Experts had examined him on multiple occasions over the years using trained interpreters. Neurology, psychiatry, neuropsychiatry, psychology, neuropsychology, all staked a claim in his presentation, tried to position him on their particular grid of meaning. But little if anything had been added to the collective understanding. Neuropsychology placed him in the first percentile on every task (the small one on the extreme left, wearing an outsized Stetson, at the opposite end from Michael). This was either because his brain was rubble, or because he couldn't make sense of the culturally alien tests, or just because he couldn't 'give a flying fuck at a rolling donut' (*Die Hard 3?*) about completing their puzzles. His injury, his mental health, his character, his imagination, his language, his culture, his room, are Teflon to our expertise. Nothing moves on. He resists everything that is offered – diaries, memory aids, planned diets, physiotherapy, computer skills, brain-injury support groups, social-skills training, meditation, yoga, drugs; and what he asks for – a drive-on lawnmower, a tiara, a holiday in Syria ('chill-ax-tastic') – is refused.

One grey Wednesday morning, not materially different to that morning when someone started writing *Crime and Punishment* or to the morning that someone else completed the genetic sequence that allowed for the production of antiretroviral drugs, an expert sat Jonah down in his living room and confronted him with the truth; hours of footage caught on the new cameras: of nobody stealing his hairdryer, of nobody eating his cheese, of nobody rifling through his suitcases or riding his baby-scooter or sleeping in his bed. Proof, belief-changing proof: a reality that must now be yielded to. Except that Jonah said that the footage had been edited: the footage didn't look like his room, but another room that had been mocked up: 'I know my room.'

This room was an impostor, intended to confuse him. His enemy was using CCTV to watch him, to know when he is out, to rob him more easily: responses that helped to earn him a new diagnosis of paranoid schizophrenia. The expert had just *discovered* another layer to his presentation, in the same way that Hernán Cortés discovered the Aztecs had Spanish flu. At my recommendation the cameras are switched off, forever, the sightless eyes of yet another herd of white elephants overlooking the fake lion and zebra on the savannah of his living room.

What else? What else can I do for him?

Dr Burns worked out of beautiful oceanside clinic rooms. The air was oxygen-rich. The ceilings high enough for a small church to fit inside. One wall of each room was made entirely of glass and therefore sunlight. I imagined that by mid-morning you might feel like an about-to-immolate insect under a delinquent boy's magnifying glass. There were framed degrees and diplomas behind his desk. A simple cross made of jetsam. A bumper sticker which read 'Follow the Bliss'. Photos of himself running into the surf, throwing his tanned blond children improbably high in the air, walking barefoot along white sand with Mrs Burns. It was the antipodes of the Formica-clad anonymity of the clinic rooms in which I worked; a shrine to personality and health.

I had driven up the Coast Highway with one of the monks to visit Burns, a fellow neuropsychologist, whose office was a few miles south of Half-Moon Bay. The monk had been experiencing sudden episodes of forgetfulness and disorientation. Burns questioned him with gentle precision, rolling with resistance, easing past defences with warmth and skill. Next he gave an informative description of the state-of-the-art procedures developed by his 'team' in the academic neuro-science department he worked out of in UC Santa Cruz. To support this he summoned a spectral rotating brain so vivid it appeared to exist independently of his laptop. Burns explained how, with new understandings of connectivity, old location-specific ideas had been replaced by distributed networks 'which speak to one another literally' (Not 'literally', I thought). He had a confection of vibrant similes: 'The brain is like a vast flowing river … like a perfectly conducted symphony orchestra … a giant colony of ants … the myriad currents of a great ocean … the ever-changing face of a beautiful child …' As he spoke different coloured networks lit up on the screen as if in synchrony. The future. Wow. It all made wonderful sense. I wanted to sit at his feet and listen. Meanwhile back in London I used a dirty tennis ball or sometimes just my fist: 'Imagine this knuckle is your memory … The brain is like major road-works on the M25 in rush hour …'

'Come check this out,' he beckoned us.

He had put on a lab coat over his Hawaiian shirt. He sat the monk on a chair in front of an interface, dabbed a little gel around his temples, then eased something that looked like a swim cap over his head, dozens of tiny leads sprouting from it like cress. Burns stood over a large mixing desk. Buttons were pressed authoritatively, faders were faded to specific settings, one earphone held to an ear like another Aphex Twin (Triplet?).

'I want you to try and move the boat to the sunset.'

On the screen before the monk there was a high-end sailing boat pointed in the direction of a Malibu-like horizon.

'That's right, no hands. Just concentrate on moving it, Brother, best you can … Relax … Just breathe … This is part of our theta wave bio-feedback training system, Dr Benjamin.'

'Right.'

'There are good early signs from the longitudinal data that it bolsters cerebral reserve.'

'Right.'

'It's moving, Ally. Look, I'm moving it,' said the monk, like a delighted seven-year-old as the boat sailed towards the horizon. One of modern life's great indignities: infantilising the old by exciting them with technology.

'Attaboy, nice work, Brother, keep going,' Burns said. Then to me: 'As well as training and a whole lot of fun, it doubles as an assessment tool by generating multi-pole quantitative electrophysiological analogues.'

'Uh-huh.'

I wonder how our patients feel when we start talking like this. Even in hyper-evolved California, 22% of the patients have sub-threshold levels of literacy.

'Just a piece of the jigsaw, but it can be a powerful tool. Such exciting time for guys like us.'

'It is,' some part of me said on my behalf. Was Burns part cyborg? Preprogrammed to mix his metaphors?

'What software packages are you guys formatted on? I could send you some templates and a couple of my recent papers.'

'Er ... Windows 97?'

I felt outmoded like a tanner or a cobbler. I was from a place, from rooms, where people used pencils and line drawings of a bishop's mitre and a wireless radio to assess a brain, where hospital canteens still served chips and processed peas, and Tracker bars were the 'healthy option'. Part of me wanted to force-feed Burns his ghost brain, sit him in his stupid animated boat and sink it, burn it, a full Viking burial along with Mrs Burns and the Burnettes. He was right on the line between bullshit and authenticity. I was on another line, between being a little bit right and being hopelessly jealous and mean-spirited in the face of his intelligence and purposefulness, the total conviction that what he was doing mattered. Right on the line; 'chalk dust', as they say at Wimbledon. I would have to award the point in his favour.

★

In another room 8,000 miles away a naked boy is slumped on a filthy stone floor. There is no electricity, little natural light. By the boy is a small plastic tub of water. With his left arm he washes his body. His right arm, which looks significantly longer, is limp like rope, his legs coil uselessly beneath him. We watch him shuffle and drag himself across the clinic to the adjacent toilet: three super-effortful minutes to make ten yards. He was born more than eight years ago in a rural village twenty kilometres away from the city, with cerebral palsy. This is his first appointment with a medical doctor. He is clearly past the point of therapeutic intervention. The father asks us when the boy might be able to walk normally. This is his only remaining son, having lost two others in infancy.

Next a five-year-old boy with no fingers and hoof-like toeless feet. His head is tiny but the forehead bosses with hydrocephalus. His illiterate mother (rates of illiteracy approach 100% here), still not nineteen, describes how she had experienced pain in her abdomen – she points to what might be her kidneys – during pregnancy. This took her to a tribal doctor in Jharkhand. He prepared a remedy for her with unknown ingredients. Her son's deformities were likely caused by this treatment. Once again there was an expectation of miraculous recovery, that twenty pristine digits were just biding their time. You might only imagine that such things were happening in different clinic rooms at the same time on the same planet, all part of a single ongoing multi-site experiment: life.

Kedar is a neuro-rehabilitation NGO situated in the east Indian state of Orissa. I had worked there in short stints over the years – long holidays, extended time off between jobs, etc. It was founded by an Italian nun, twenty-five years ago, having observed a terrible shortfall in the local provision for children with neurological diseases. Dr Brossard, a French neurologist, was an early recruit. He more than anyone, through his understated, patient brilliance, had been responsible for integrating Western standards of professionalism with a rural culture that was only just beginning to understand the necessity of using rigorous empirical methods to help with certain aspects of being alive. The enterprise had grown steadily from a room above a *dhaba* in the city of Bhubaneswar to what is now a purpose-built village covering 100 acres of farmland with 400 children residing, some with their families living with them, and 250 staff, many of whom had arrived as child patients and had gone on, in basic wheelchairs or outsized crutches, to train in therapeutic specialties, or in the case of more significant disabilities, to more menial work in the village. In Kedar the patients become 'doctors'; at home in the UK the doctors live in unacknowledged terror of becoming patients.

Though underfunded, of questionable skill in aspects of clinical work, often vague, always chaotic, struggling to organise themselves along Dr Brossard's guidelines for quality of care, there was an easy optimism among the staff, mirroring that of the children, making the many different healthcare placements and settings in which I had worked in the UK look institutionally depressed by comparison. They advocated most of the same values – dedication, care, respect, compassion – as the Trusts at home, but here they grew out of experience and belief, rather than being branded onto it by an expensively subcontracted management consultancy.

This despite what were unimagined forms of suffering. The final patient in the morning's clinic is a nine-year-old girl who has Sanfillipo syndrome, a rare regressive disorder. She is carried in and lain lifelessly on the examination table by parents who are impoverished labourers. The girl has no neck control, cannot see or hear or speak. She has up to ten grand mal seizures a day because her parents can no longer afford to buy the drugs that are required to control her epilepsy. Two years ago she was a normally developing seven-year-old. The disease, which has a mean mortality in the early teens, takes away each faculty in the reverse order it was acquired: first complex language, then noun phrases, then words; subtle then gross motor function – neck, then trunk, then bowel control – until, just babbling and writhing, until nothing, a regression with death replacing birth, the cruellest imaginable parable on the construction and deconstruction of functioning, identity, parent–child attachment. In their brief consultation Dr Brossard would notice how much the parents believed, despite any explicit evidence of her responsiveness, that their child felt the force of their care, felt loved at least. And he gently supported them in this belief, priest more than doctor in that moment. There is nothing else to be done. The father who had no money to offer for today's consultation or anti-epileptic drugs would, after the morning appointments were finished, carefully sweep and scrub the clinic room clean; an offering of his deep appreciation. There on the wall the Kedar Calendar, each month the picture of a smiling deformed child above a quotation:

'We are never more than one grateful thought away from Peace of Heart.'

'The moment one definitely commits oneself, then Providence moves too.'

'Joy is the happiness that doesn't depend on what happens.'

Sentiment: a lingua franca, or just the limit of language. I think of my cohort, of us: over-educated, pleasure-bound, barrier-nursed by our rarefied culture, too ironic to let basic values survive our scrutiny, condemned because of it. In private we might want to believe, but we fear it is too late. Only here, these things still mean something authentic.

'What else?'

We were no longer in Bede's room but walking down the driveway, high above the coastal road. It was late in the day. There air was still. Bede looked tired, he'd already been up fifteen hours I suppose. The pressure on his speech had relieved itself a little. These subtle changes gave him a different quality; pensive, elegiac even, as if the mania were setting in him like the evening sun.

What else?

He had fallen in love and now the affair was approaching its end. It can be easy to lose faith, forget what moves us towards something in the first place – the seeds – a kind of autobiographical amnesia for what inspired us, a subtype of Benjamin's syndrome. It was as though growing up had left no rings in Bede; we can lose even the most profound, hard-earned wisdom through lack of care. Hence the need to retrace our steps on occasion, rediscover a treasure in a story of origins, even if we've made it up.

What else? I think about deflecting but I don't ... I remember how still and innocent Bede looked as I finished telling him the story of Lotte, another love story, and a story I often struggle to tell – often forget to remember – of how she guided me to do what I do now. Way below us the sea heaved and the kelp bobbed with it, looking like the heads of so many men. I envied his heartbreak. I envied his untrammelled future.

Then he told me he wanted my job.

That was pretty much it, time for me to move on. Life moves through different rooms, a series of conversations – monologues, prayers, group discussions, but one-on-ones mainly – in which we try to change one another or ourselves, even try to change places with one another. We turn the conversations into stories, stories with endings, when there are no such endings, because there are no such stories, just people talking.

A few weeks later I received a letter from the prior informing me that Bede had left the monastery. He returned to the UK, moved into a small bedsit on a housing estate in East London to study for his new career at a nearby community college. These days he ran along the canal with its toy-town housing, dirty-looking swans, furious cycle commuters or pedestrians walking homicidal Staffies. No more time for sitting still: evenings and weekends he worked as a rehabilitation 'technician', changing incontinence pads for unreachable stroke patients, hoisting quadraplegics from bed to bath to armchair to bed again. After a couple of years he got a basic degree, a year later a master's, several years after that a doctorate, and then another one just for good measure. He stopped going to church, forgot all but a few scraps of the psalms ('Confuse, O Lord, divide their tongues, For I have seen violence and strife in the city'), met someone, moved in within a fortnight to a spacious, light-filled apartment, more than he hoped for. He started going to AA again. Within a year they had a son, or a daughter, which he celebrated. He stopped AA. He had another child, couldn't stop celebrating, split from the mother, moved back into a bedsit, met another girl, stopped going to AA, ran to exhaust himself, started meditating then stopped again, strayed into church once and left before Eucharist ... Twelve years after paradise he began his clinical work.

There: another story. Really, though, I don't know what he did. Just as he had fantasised about my life, I would do the same with his from time to time. Until, that is, I forgot all about him.

Tracy

Tracy walks into the afternoon follow-up clinic, smiling with a loose mouth, steps uneven, leaning to her right side, gait too wide, flanked closely by both parents who are wary of a fall. They have travelled all the way from Aberdeen for this clinic, twelve months after the accident. People often wrongly assume that the younger the injured person is, the better; it fits our sense of the early powers of recovery, of burgeoning, unstoppable life. But in the case of paediatric injury, despite the prodigiousness of plasticity in the young, often the only growth is in what they lack. Windows for cognitive development are missed, the lag behind developmental milestones grows further and further, so that by the time the infant reaches high school the difference between impaired and normal functioning is often unbridgeable.

The cause had been meticulously documented in the notes, like a forensic report; I wasn't going to ask them to go through it again. The father had fitted a seventy-inch wafer-thin television to the master-bedroom wall of their rent-to-buy new-build, three-inch rawl plugs drilled into MDF panelling, so that it faced the super-king-sized bed whose mattress used 'lunar technology' for comfort. This is how we look after ourselves. Unfortunately the 'on' button, being 120 cm off the ground, was just within reach of the outstretched arm of Tracy, their four-year-old daughter, who measured a little less than a metre on the last height chart. The thin wall couldn't support the fittings. Tracy's mother found her convulsing body on the floor, obscured except for her feet and hands by the massive fallen screen. An image from a cartoon, with Jeremy Kyle playing on top of her. Daughter, mother, father, marriage, family, broken in an instant, by an outsized television. No need to tell *them* their family will never be the same again.

Tracy is, I think, trying to tell me about a princess. Her speech is so slurred that I can't understand it without her mother interpreting; her world is cocooning, already her parents have become her only non-professional ciphers. I notice the way the father defers to his wife, as though still apologising, and how she couldn't be more sealed off from him, and from me. She concentrates all her attention on her daughter, no longer expecting anything, attending yet another clinic at another hospital because that is what she does these days. No more questions about prognosis, where she might be in two years' time. Everything has been decided.

Tracy has climbed onto my knee. I think she is saying, 'Nobody's perfect,' in slurred sing-song Scottish, pulling my tie like a toilet chain.

'Hannah Montana fan,' says the father.

'Are you OK, Dr Benjamin?'

I am not OK. It's one of only two or three occasions in my career where I have cried openly, momentarily unable to keep myself at bay, to at least give the impression that I am managing my feelings.

'Yes. Please excuse me.' Tracy's father moves the box of Trust tissues towards me.

A television devastates a child's brain. These things actually happen, once in a decade maybe, across the entire population. There is no Literature on this to consult, but if you can imagine it, and it's in keeping with the laws of physics, even if the incidence is 1 in 10^8, then it's going to happen at some point. That is the medical cosmology.

What I don't tell them is that Tracy isn't the reason I was crying.

Bronwen, my eldest daughter, was not quite two. I was still living with her, mistakable for any one of the million ordinary-looking millionaires of London's Zone 2 – slowly getting the hang of being a father – but not for much longer. Earlier that day, the same day that Tracy would end up pulling my tie, Bron had been sitting in a high chair at the table, playing with her breakfast of chopped banana and apple. I would normally have been at work. At the time I was based on the neuro-surgical ward of a children's hospital; 'Toucan' was for children with tumours, TBIs, cerebral viruses, intractable epilepsy amongst other things. Inevitably the children and their families would toggle between Toucan and Neuro-intensive Care (no cuddly animal name for that ward), depending on how things went. There was a soft thud as Bronwen's head hit the high chair's wooden table, where it remained, unmoving.

My partner Helen and I had a brief stand-off about calling an ambulance. Our second daughter Cordelia had been born a few months earlier. We had gone back and forth about which hospital was the nearest (the Royal Free in Hampstead by 0.4 miles) before deciding on Queen Charlotte's next to the Hammersmith because the maternity suite was brand new: we would be part of *its* christening. In fact, having two hospitals close by wasn't reassuring; it made us feel between things.

Less than two hours before Cordelia appeared and Helen still wasn't sure if she was having contractions. When I suggested she might be, she didn't think so. When a little later she thought she was, I questioned it. We often had difficulties communicating, but one of the things I liked about her was that my doctorliness earned me no extramural authority. It was one of the things I didn't like about her as well. She persuaded me to bring the car round to the front of the house. By the time I got back the arguement had moved on…

On that occasion the ambulance didn't arrive for an hour. And then, two arrived at once, one from the Royal Free, another from Charlotte's. By which time Helen was sitting on a beanbag with the formally perfect 'Betty' (I had to fight long and hard for her not to be 'Betty' or 'Rita' or 'Ivy': war names, the names of dinner ladies at my northern school, become West London chic), both of them so insouciant, so 'over it' – the umbilical cord which still connected them resembling an exotic Japanese fashion accessory – together, a mini-Russian doll disgorged from its larger mother; one might expect Cordelia to be holding her own, proportionately smaller, daughter. Under instruction, I used a cake slice to scoop the placenta into a freezer bag, for resale in the local farmer's market. (The next day several bouquets of flowers arrived, congratulating *me* on the delivery: 'Well done, Doc!', 'Safe pair of hands.' I had done next to nothing, as Helen reminded me. She knew how doctors would always be the central characters in their own emergencies, the patients just extras.)

Unlike Bron's birth – and my own, for that matter – Cordelia had torpedoed into the world, maybe too quickly; she was silent, blue, shocked-looking, breathless, meaning *not breathing*. I turned her upside down and thumped her back, a little too hard, as I'd seen them do on an episode of *M*A*S*H* as a child. *M*A*S*H* was the only medical drama I followed, and that was decades ago. I had therefore missed out on the updates to my Continuing Professional Development that *Casualty*, *ER*, and *24 Hours in A&E* would have provided, my emergency knowledge suspended in gross approximations of military hospitals from the Korean War. Nevertheless thirty seconds later she coughed and cried–in her first inhalation.

An hour later and six or seven ambulance folk were sitting round the kitchen table like bank robbers, drinking tea, chatting about annual leave allowances and basement excavations, while in the next room Helen, Bron and I waited for another hour for a midwife to arrive and cut the cord.

Therefore this time, with Bron still face down in her breakfast, we decided not to call an ambulance. Helen drove the 2.8 miles to the Royal Free in silence. In the back seat I did what I could to rouse Bron; I shouted, shook, splashed bottled water in her face, shook her again, harder and harder. I dug my nails into the soles of her feet − *M*A*S*H* again − beads of blood emerged where my uncut nails had pierced her. Even then, even when I've never been more frightened and sick and absorbed by anything in my entire life, there was still enough rogue bandwidth to wonder if these fresh stigmata would trigger the Child Protection procedures at A&E, and then to wonder that the modern mind given unimaginable security and freedom would still gravitate to the Magnetic North of risk. Nothing stirred her, she was cold.

Helen had taken a rat run to the hospital but it was the time of the school drop-off and Arkwright Road had a long queue back from the top of the hill. At one point an ambulance came the other way past the line of stationary cars. I got out and waved it to stop as it sped past, *giving back nothing*. At times of dreadful stress cortisol stops the conscious memory system from laying down new episodes. The next memory I have is getting out on a zebra crossing at the top of Pond Street. I put Bron under my arm like a deflated rugby ball and ran down the hill through the traffic. It wasn't a long way, 500 yards at the most to the main hospital building, like interval sprints on my Run Programme these days, but I wasn't so fit back then. I remember looking for signs to A&E. I wanted to scream but I didn't. I would have known where to go straight away if it were my hospital. Instead I was lost in what seemed to be the subterranean space of a multistorey car park, like something out of a post-apocalyptic dream; the landscape of a terrible obsolesence. By what rule does the architecture, anatomy and efficiency of a hospital coincide? By whose imagination were emergencies buried deep underground? I wanted to scream but I couldn't. Nor did I dare look at the cold weight wedged under my arm, more deflated now, like wet swim kit.

I was still at home that morning because we'd had an argument. The same argument as always, but the accent was different because it was the end. The night before we'd watched a box set together. The friend who had lent it said we'd love it, that it was cheaper than marriage guidance. It was quite well done. By the time we got to the third episode we were sitting on the same sofa, Helen resting her head on my shoulder. The kiss goodnight had real feeling, an umbilical cord to a time before everything that had happened. It must have been about midnight, way past our bedtime. I watched one more on my own, a 'quick' one; it wasn't *that* well done. I watched another. It was starting to annoy me. I watched another, and another, and just one more until Helen came downstairs with Cordelia and Bron; it was 6.15 a.m., time for breakfast. I was on the second episode of series two.

I stayed home to help with breakfast because I knew that it was over.

'Where we are, Daddy?'

...Just as we made it to the reception desk; I nearly dropped her. Time started again. A triage nurse was asking for details. In a moment the crisis had gone from being self-evident – a dead or near-dead toddler – to its opposite. I sent out a profusion of meaningless-sounding urgent words, trying to recount every moment of the last thirty minutes in a litigiously scrupulous sequence with a healthy-looking two-year-old on my shoulders. And as a consequence, I felt with each passing moment the increasing pressure of persuading the nurse to take me seriously, a delicate balancing act for all patients: too much urgency and run the risk of appearing bullying, hypochondriacal, histrionic; not enough and you don't even register. The art of being a patient is saying the right things in the right way – not too expert, not too lay – to trigger the urgency in them, the professionals, the arbiters. The first part of the deal is that you have to contain yourself, turn the volume down until they can hear your story, until they can tell you your story.

In the background Bron was wittering:

'Daddy? ... Daddy-doctor? ... Daddy-lunch? ... Daddy-oh?'

And on she went, reassuringly, throughout the two-hour wait to be seen.

'Axdent *and* Mergency ... ? Salt *and* vinegah ... ? Mummy *and* Daddy ... ? Scooby Dooby Doooo.'

The on-call paediatrician checked her over rather languidly I thought. I considered telling him what I did, where I worked, who I worked with. I thought about sharing ideas of possible differentials, necessary investigations. I did nothing. He stared deep into her big brown eyes, shone a light there, made her go 'AGGGHHHHHH', listened to her chest ('tickles') asked her to track his finger. The wittering had stopped, she was a descended angel: bashful, polite, radiant, stoic, getting back what she gave out in smiles and winks. Coy behaviour begins early in development to elicit nurturance and reduce the possibility of adult aggression. Together now in each other's arms, doctor and patient: the image of health, according to him.

'It was a non-febrile convulsion. One-off, most likely. She's fine.'

Bron looked fine.

'Most likely?'

'1 in 10 to the 2 it's something else.'

'You mean a thousand?'

'No, I mean a hundred.'

Doctors sometimes prefer numbers to words. They lend veneer to the meticulous probabilistic reasoning that – you (and now me) like to think – has underpinned your care.

'You mean it just happened for no reason.'

'No reason we're concerned about.'

Part of us prefers some cause to none, bad news to no news. He smiled, winked at her once more. Bron beamed back, looked at her toes, then turned to me sheepishly, nuzzled in my chest. My daughter – the actress, the hysteric, the malingerer.

I'd been awake for a day and a half. The follow-up clinic that afternoon had just finished. After Tracy, I had assessed a dropped baby, a skateboard fall, and finally a road traffic accident in which the young TBI's father and brother had both died ... Unceasing suffering, death, despair compressed into less than two hours.

Walking absently through busy corridors I bumped into Francis, one of the paediatric neurosurgeons, in the corridor outside theatre, taking a quick breather from a long operation. He was a gentle, kindly, introverted man, not altogether typical of neurosurgeons in my experience. He'd had twins in his late forties who were only a little older than Bronwen, his wife knew Helen from nursery. Without intending to I found myself telling him about what had happened that morning, an anecdote without a punchline, another subtype of Benjamin's syndrome, a story that turns into a question and therefore a referral, because in telling it you are struck by the fear that the ending hasn't yet happened.

'How long was she unconscious? Glasgow Coma score? Signs of status?' And more questions, all the while his courteous smile lightly withering.

'I would take her back in, Ally,' he said softly but insistently, 'make sure she has a scan and a lumbar puncture. These guys are good but they don't always pick up on some of the things we see here.'

Time stopped again. The second nauseating dash of the day: 4.3 miles back to the Royal Free on my bike where I would meet Helen and Bron, who had spent that afternoon at home eating sweets under a blanket in front of *Scooby Doo* (one day in sick role, 'Doctor's orders' – wink). I remember nothing of that effortless, motionless ride; no traffic, no pedestrians, no lights, no stopping, all attentional resources elsewhere or nowhere, like a vipassana meditation where the entirety of the phenomenal world is annihilated. Not quite the entirety; still the stray mutinous thought – too much bandwidth again – that this emergency was something I had helped to create post-argument, to show my worth as a father.

'I'm not going to do a lumbar puncture on her.'

It was the same paediatrician; I was ready to grab hold of his tie now. Bron was in my arms, batting her big eyes, trying to look ill at the same time.

Sometimes it's a straight fight: doctor vs patient. There are other metaphors besides war (it didn't feel *metaphoric* at that point): paternalism; the machine; a theatrical two-hander; increasingly the marketplace – metaphors which do much to reify the relationship. Whatever the rhetorical figure, some form of asymmetry is at the heart of the medical enterprise, however 'patient-centred' or touchy-feely the approach. The lopsidedness is co-constructed: it's in the patient's interest to be told they are ill if they report it. The trick is how to transform your experience of disease into medically salient descriptions, make this hospital visit meaningful. And of course the doctor who becomes a patient should be expert in this, uniquely able to negotiate the asymmetry from both sides, skilfully representing his daughter's symptoms in the most lucid, un-histrionic, medical-sounding language:

'Yes you fucking are going to give her a lumbar puncture, *pal*—'

'I'm not and don't threaten me again,' he interrupts, 'and I'm certainly not going to do an MRI on her. She has no symptoms or history that would warrant it. Besides, it's not good for kids her age. Your friend at—'

'Mr Thomson, my *colleague*, is an experienced neurosurgeon at a national hospital.'

'Exactly, all he sees are exotic high- and low-grade neoplasms which may or may not have symptomatic seizures associated with them. His experience is skewed, his sample is not *normal*.'

This had been the least normal day in my experience so far. But it was the last gasps of family life more than the suffering of others (including my daughter) which pricked my tears in the clinic room before Tracy and her parents, because with that failure came the familiar thought that I would miss, have missed over the intervening years, all those episodes – from terror to joy and everything in between – in the lives of the girls, and with them the chance to save, protect, influence, accompany, hold back or worse: to be a father in other words, futile though the instinct often was, and to be nurtured in turn by the experiences that go with it. I scarcely understood the freight of those tears then, worry it still may not be fully apprehended, that the grief is still maturing. It was Family Benjamin's turn to never be the same again.

The paediatrician was right: Francis moved in rarefied waters, his bread and butter were the horrific and rare; the one in a million was commonplace to him. In statistical terms Francis, primed by his global sample (work took him across the world and brought the world to him), was on red alert for a false negative: the chance of deeming a child healthy when he was ill, potentially forgoing urgent, life-saving treatment. That didn't much concern the paediatrician, who sifting the relatively tiny population around the hospital was more concerned about false positives; wrongly diagnosing a well child, which hypo-chondriacal Hampstead parents were always willing him to do, and the collateral risk of administering unnecessary, even harmful treatments. There was always the danger of these two errors, the possibility of over- or under-filtering, even when the most advanced paradigms that Californian artificial intelligence had to offer were used, even when everything has turned to numbers: you can't eliminate the *wetware*.

For now the numbers were the paediatrician's concern alone, and we were safely in his explanation, snug in our paternalist metaphor. And Bron certainly looked fine: there was no cognitive dissonance there. Equally, an hour earlier in the corridor outside theatre, I had felt the horrible click of a different reality falling into place as Francis hypothesised an underlying tumour. He had clearly said that he would scan if it was his daughter, and though his style was less bludgeoning, he spoke with the same conviction as the doctor facing us now, telling us that he wouldn't scan if she was his.

The 'my daughter' trope: do these men really know what they are saying? I blame *24 Hours in A&E*; the for-the-cameras industry patois, the speak-for-itself doctorly concern. But relations don't speak for themselves, there is no norm we can take for granted, not between doctors and patients, nor between fathers and daughters: take Shipman, take Lear. It could be that one doctor doesn't care so much for his daughter, while the other is morbidly codependent, one daughter might just be more difficult to love. So which one is it?

'Can you see why it might be difficult to accept your reassurance?' I say it as soberly as possible; inside I'm in desperate, abject petition.

'Yes, and it doesn't mean I'm not right.'

We weren't the first people to be caught between hospitals, between differentials, between numbers, between two more or less sympathetic conscientious doctors sampling probabilistically between two diametrically opposed realities, between *their* daughters in other words. Sooner or later the truth would emerge, probably. For now we were on its edge, both sides perfectly plausible, both somehow fitting with our sense of what we thought or feared or hoped for.

As you were: we choose health.

Bron was on the doctor's knee, as Tracy had been on mine a few hours before. We were a family again, the five of us: me, Helen, Bronwen, Cordelia and the on-call paediatrician who had *saved* our daughter's life – six if you include his own daughter whom he loves so dearly. Francis and his uncared-for child were forgotten. As were Tracy and her family.

'Try and put the whole thing to the back of your mind,' the doctor was telling us.

'Already done. No problem. Like I have a choice,' I thought but didn't say.

'You look shattered,' the doctor told me, on the other side of confrontation. For the first time I felt how tired I was after no sleep the night before. 'It's tough with two under-threes, but it gets easier,' said the doctor. I imagined Helen's teeth grinding.

'What if she has a second episode?' I ask.

'That changes the incidence rates significantly,' he replied, meaning that Francis and the daughter whom we had implicitly decided he loved only half-heartedly would have been right all along.

We were family Benjamin again. But not for long. Our doctor remained in the hospital when we went home. A few weeks later, I moved into a friend's spare room. Like Tracy's father would, I imagined, or already had to all intents and purposes. I didn't know what the end would mean then, of course; I might have tried to stay longer if I had. But at least – and even now, nearly ten years later as I write, tears are choked off by a shiver of relief – my daughter would remain safe in her home, along with her sister, her mother and her undamaged brain.

'L'

'We've thought long and hard about it: all things considered we think it best that names are changed.'

'That's bonkers: he's not some random case, everybody knows him,' I say.

'We really have given this a lot of careful thought,' says the psychologist convening a new, different type of meeting, at which I am due to give a short talk about 'L'.

'He's not even a patient, he's a colleague. Which "we" … ?'

'He *was* a patient,' he says. 'His wife and family have a right to confidentiality.'

'Everybody will know it's him.'

'Holding back his name won't stop you from telling his story in a way that touches people,' he says.

'It won't be *his* story.'

Case studies deaden, take life, rather than give it. I wanted more for L. So typical of clinical psychology; name-erasure alongside other more or less deft effacements of human interest, little technologies to dry things up. (One particularly baffling policy was the surreal use of the word 'apple' to obscure sensitive information on departmental reports.)

'We thought we might close with a brief meditation.' His voice is breathy, concerned, like a West Coast podcast. And that 'we' again: he means his profession, the sympathetic in general, the softly spoken, the mindful, the sane: come on in, the water's lovely. I felt my chest tightening. 90% of clinical psychologists – even the men – are white, middle-class women. Sometimes I felt emasculated. On the other side, the doctors, the managers, the accountants; hyper-busy boys who can't make eye contact. Sometimes I felt e-feminated. It was far from what I had hoped for, all those years ago when Lotte had first sown the idea in my mind.

I had a ten-minute walk from Marble Arch. It was a second date, the first having gone unusually well. On such occasions I always invited them to the Greek off Edgware Road. It had a seductive mix of casual *grande bouffe*: each of the ten dishes on the menu was carelessly brilliant; a scruffy interior – you might find a sock or a toothbrush under your chair – a whopping vinyl soul collection, bits of *Star Wars* figures, Rizla packets, disused circuit boards, like you'd walked into a student's digs and he was an ageing DJ-stoner genius chef, when he wasn't gaming, mouth ajar, in full view of his customers. This Greek knew me from previous dates, helped me where he could with gratuities, cheek-pinching, double kisses, the feeling of singling me out from the rest, a fellow exotic, a Mediterranean brother, etc. In truth the Greek was a Cypriot raised in Penge. In truth there had been far too many awkward, depressing nights spent here in the years since I moved out of Helen's home. I called them dates, but it was obvious to anyone that I was really conducting interviews for a kidnap, screening for possible detainees to jump-start a new family with me at its centre to stop me feeling the full weight of losing my own. Obvious to everyone but me. But this date was different, it felt moderately interesting. That's a lie: I was excited. She was from Chicago, sharp but not just *sharp*, soulful also, a Boo-dhist, the two seemingly syncopated. There was something attractively dark about her timing. On the first date she had compared looking for a guy at her age (at 'our age', give or take a few years) as like looking for a parking space at Waitrose on a Sunday: 'Only the disabled spots weren't already taken.' She had me.

I called L en route; an autumn evening in 2016. He was the person I was closest to at work. I was grateful for the chance to hear myself describe her to him, burn off a little of the excitement which might otherwise capsize me one way or another.

I still regret not letting him speak first. After a characteristic pause he said: 'You're on fire. What's it going to be: water or petrol?'

Finally, a full five minutes into the conversation, I asked him how he was. It had been several weeks since we'd last spoken, on a late summer cycle in the Surrey Hills in the days before my Stalinist Cycle Programme, when I could still ride for the sheer, creaturely pleasure of it.

And then he told me.

The date was a disaster. She offered to rearrange but I never saw her again.

L was a social worker, specialising in neurology. These are the people who pick up the pieces for the patient and the family after the life-changing diagnosis or trauma, organising often complex and nuanced packages of care after discharge, either in adapted home environments or specialist placements. They often get involved in decision-making around supporting the patient's capacity to function, and in the absence of adequate mental capacity, helping to make decisions in their best interests. Such issues are the most messy, intractable but critical elements in a person's quality of life once the medical position has clarified. Most of us run a mile from the mess, however much our expertise is required, and, typically, expect the social worker to resolve it quickly and bloodlessly. L was brilliant at it. But that wasn't why I was drawn to him, the reasons were much more visceral.

He was a few years older than me, a sixth-former to my third-year. He knew about music – 'dirty rockabilly' his favourite – was talented at sports, wore buckled winkle-pickers to work, still ate fish and chips, supported Fulham, liked Russian novels and French films, and like them didn't talk much except to be wry, offbeat and – I might as well have been a teenage girl drawing his picture in biro on my maths folder – to my mind he was about as unlike a doctor as it was possible to be. We worked on cases together in post-acute neuro-rehabilitation, where acquired brain injuries, once stabilised, would get time-limited, multidisciplinary interventions. Though angular, tricky even, he brought unusual interest and therefore depth to each case; he was able to think about the meaning of injury, imagine for himself the lived experience of the person, their family, how it would translate into their forever-changed lives. It sounds so simple but few want to do it, and fewer still can without getting snared by their own preoccupations. L made cases for patients, made them living terms in a system where otherwise they might have failed to register.

The door of the lecture theatre swung shut, and out of the corner of my eye I noticed my line manager sneak in and make her way to the back. Even she had a soft spot for L.

'If he sounds earnest or soft-headed it's my fault, you know he's not like that. He was always reminding me that somebody in this giant ugly place has to see the angles: [impersonating his cockney accent] "I'm talkin' about the fuckin' socio-economics of head trauma – you think it's a coincidence half of them are Eastern European manual labourers? Are you sure? Pull the other one ... " It sounded funny in his voice, like a punter at Walthamstow dog track had swallowed *The Dictionary of Marxist Epidemiology.* He wanted us to think beyond the confines of our clinical training, like how inequality might account for a lot of the variance in medical tragedies; in a dozen years of training I'd had one seminar on that possibility. But this wasn't really about formal learning; he'd had a life before this work; a lorry driver, an activist, a nightclub owner, a failed artist. Not to say some doctors don't have breadth; I know a few with a scarcely credible range of interests. But the times are changing. I recently did some research using a control group of juniors. On a brief general-knowledge screen most of them didn't know who Martin Luther King was, or who wrote *Dr Jekyll and Mr Hyde*, or what the minimum wage stood at ... "Big deal," you might say. "As long as they tell me what's wrong." But what if in cultivating super-specialists we are inadvertently hurting ourselves? What if their narrowness goes beyond general knowledge? Do you see where I'm heading? You don't? What I'm trying to say, what he was trying to say, is that if various basic aspects of our world are undiscovered, unnamed, incomprehensible, how can we really detect or diagnose or treat the unknown or the strange-to-us in our patients?'

I paused, and took a sip of water.

'Now Lewis was no intellectual, let's face it ... Oh fuck! Sorry. Well there goes his anonymity, like you didn't know ... Anyway "L" – *Lewis* ...'

I could feel the convening psychologist looking at me from behind the lighting desk.

'It was a slip,' I imagined saying to him.

'Now who's being naive?' I imagined him replying, lips thinning.

'Fuck him,' I hear Lewis say. He had come to life: it was *his* story now, even if it wasn't.

'... I could say he cared, but what does that mean? It hardly makes it past our ears any more; we've all heard it so many times, in such disfiguring ways. So I'll put it another way: he was the one person I've met in this entire hospital who seemed to have time, in the way they say of swamis or good batsmen, even when he didn't. It's a mosh pit here most days: never mind the patients, we batter each other with anxiety-ridden updates about how critical the situation is, "from *my* professional position", forgetting who or what is really at stake. Or we put on grave, tight, public faces and garble urgent information for the patient or their family, euphemising it to death usually, or spilling its guts in a few thoughtless, torrential moments, leaving them confused and terrified.

'Not Lewis. However dire the situation, however under the cosh he found himself, he could take the heat out, play it with soft hands – I saw him bat once, he could have been a first-class cricketer. The super-uptight became more exasperated, that he wasn't answering them right then and there. But for most of us, his manner made us take a breath, separate just enough to see what was really going on. And he spoke to you in exactly the same quiet, matter-of-fact tone whether you were the surgeon general or the unemployed father of a pregnant teenage crack-coma.

'But that's not, but that's ... excuse me ... and ... but ... so ... and.'

Is what I said in the first three and a half minutes of the Schwartz Round to 200 or so colleagues, in May 2018 with summer swallowing spring, as it did these days. Schwartz Rounds were a recent initiative copied from the States: hospital staff voluntarily meet in their lunch hour to talk about topics that don't find time in the usual run of things – accidents, malpractices, ethical quandaries, staff trauma – exactly the sort of discussion that Lewis would have advocated, though he would have probably thrown in something about horses and theatre doors as far as our Trust was concerned. When I say 'lunch hour', nobody had an hour, most never ate lunch. So it said something about the psycho-spiritual hunger of the staff that there were this many of them packed in here.

The situation in the hospital had deteriorated badly. A few weeks before, a new CEO, looking all of seventeen years of age, emailed the entire 10,000 staff to say that the Trust had gargantuan debt. Extreme cuts would have to be made, jobs would be lost, jobs that were often required to pay off personal debt. A few days before the Schwartz Round another email told us that the Care Quality Commission (CQC – the independent healthcare regulator) would make an emergency review of the hospital, so concerned were they about clinical standards. To do this they would have to infiltrate us without disturbance, disguised as surgeons, nurses, technicians, porters, patients even, among us now; that was the rumour anyway. The general hum of insecurity was felt in the auditorium.

The topic I and two other speakers had been given was to describe, in no more than five minutes, a colleague who had a lasting effect on our careers. The convening psychologist had already vetted our stories, at least as far as we knew them. But it wasn't until I was three and a half minutes in that I remembered this little hymn to my friend, who was known to all, was in fact a eulogy. Start a story like it's a jaunty anecdote, forgetting it's a vomit-inducing catastrophe: classic Benjamin's syndrome. I'd been momentarily transported by my own voice into Lewis's company, and I tried to keep it there, suspended in his presence, three-quarters of the way through my allocated time (the psychologist was a stickler), stretching the moment out indefinitely: a manifesting denial, a strange loop, as though the longer I was stuck in his life, the longer I preserved it, the longer his death was postponed – that's what stories can do – knowing that to continue would mean heading towards an ending that was about as bad as things could end, with him, dragging his skeletal frame … no: some details really should remain confidential. From diagnosis to death in little over sixteen months, taking everything he had – voice, swallow, muscle, mind, breath – in terrible, quick succession. And it wasn't even that which I shied from. I was stalling because I had to kill him quickly, in half a minute at the most, and then, if I had the guts, speak about the other, less obvious, less hortatory, more profound effect he'd had on my career.

But I can't kill him. Stuck on 'but', stuck on 'and'.

Twenty-five years earlier I remember running to the exam hall to hand in a bulky thesis, nearly a hundred pages, tens of thousands of words, perfected over what felt like years. 11:57 a.m. I have three minutes to get there. I am running down the middle of the road dressed in a dinner suit and white bow tie, a gown and a mortar board. There are crowds of people on the high street, an audience with cameras for my dash, but really Japanese and American tourists. I skip up the stairs of the hall where an administrator meets me and asks me to sign a receipt, handing over my tome as the clock strikes twelve, and as the faceless mandarin relieves me of my masterpiece, I notice the last unpunctuated word of the last sentence of the last paragraph: 'and'.

Now, for the second time in my professional life, but this time in front of 200 paid carers, unguarded tears were streaming down my face. Once again their cause would be misread.

I imagine crying for my remaining allotted time and then crying some more, so that the two other speakers – a

senior anaesthetist and a clinical governance manager –
wouldn't get the chance to speak about their heroes, the
proposed meditation cancelled, until someone would come
to console me from the audience, offer me tissues and soft,
silencing words; only, Canute-like, I would keep on crying,
as though energised by the display of grief, as though channelling a capaciously lunged Hindu widow. The consultants
would be the first to leave – allergic to tears, to disinhibition
in general – followed by the managers, the auditors, security,
until there would be twenty or thirty people crowding me
with their understanding, 'uh-huh-ing', nodding like dande-
lions in car exhausts, arms round me, stroking my hair,
telling soothingly embarrassing stories of their own. But to
no effect. On I go, in giant gulping howls, nose streaming,
capillaries bursting – the grief's origins totally disguised –
for five, ten, fifteen minutes, clinical levels of emotional
non-governance, until all the gathered hearts are baffled, or
stuck, or angry, or just uninterested; until only cleaners and
porters are left; until one by one they make their excuses,
leaving me and one other straggler in the hall, the last person
but one; an outpatient looking for his appointment, a stray
from psychiatry, the undercover man from the CQC, unsure
whether to turn the light off; until alone. Then the truth
can begin.

It wasn't Lewis I was describing but how I imagined him, the
story of how I needed him to be. Lewis himself was a little
different. Just as there were things I made up that I did tell
the Schwartz Round, there were other things that actually
happened that I wouldn't tell the audience because I wasn't
sure, in their well-meant piety, they could handle deviations
from conventional hagiography. I didn't tell them how
sometimes we would bunk off to local curry houses where,
both up against it with child payments and rent, we would

dream up moneymaking schemes over half a dozen pints of lime and soda. One such idea was for a calendar or diary which would be eighty-odd years long rather than the usual annual. As you came to the end of each year the diary would tell you the median age of onset for all the horrific conditions you had just exceeded. So on your eighteenth birthday for example you could say goodbye, with some statistical confidence, to juvenile-onset epilepsy, Sanfillipo's, malignant tumours of the posterior fossa. But before you cracked open the champagne the diary would introduce you to the diseases into whose orbit you were chronologically and statistically heading: multiple sclerosis, fronto-temporal dementia, Parkinson's. As health falls off some sort of cliff at around seventy, the biblical span (only 5% of the UK population present at their GPs with so called serious conditions, by seventy-four it's up to 50%) the pages would suddenly become thick with new items, until the very last page, which was left shockingly blank … Our kidneys were green fruit pastels by now, but we were on a roll: it would need to be well made and boutiquely packaged; for sale on the South Bank Christmas market perhaps, retailing at £25, the venders wearing scrubs, stethoscopes, etc. Really we were lacking any entrepreneurial nous or intent. Without acknowledging it as such, this was our playful, puerile way of keeping the things we encountered every day at arm's length.

In the calendar's terms Lewis was an outlier. The onset of his disease was several years below the mean. Though we spoke on the phone intermittently during those last sixteen months – while he still could – I saw him only twice. The first time was at the Greek, Christmas 2016: terminal Lewis, our last date. I hoped a huge meal would help him put some weight back on. On the surface he was his usual laconic self, but under that a desperate wilfulness – which shrilled as the weeks passed – that his fitness and character would lend him more time. We'd cycled a hundred miles in six hours only a few months before;

surely that would count for something? He spoke of us heading back out into the Surrey Hills on our bikes; he spoke of completing his research in cross-cultural understandings of neurological illness; he spoke of a patient with the same diagnosis under the national specialists at King's who had gone symptom-free for nearly a decade. But in the gaps between what he said, the disease – a disease which more than any other makes grown neurologists wince – was working: his food remained largely untouched, coffee leaked from the corner of his mouth, his speech staggered drunkenly, normally a sharp dresser his shirt flapped loose at the back, his shoes remained unbuckled; a loss of power and dexterity in his left hand which made such tasks difficult. Then he put his fork down, this deeply sane and faithful man, and asked me with a straight face what I thought of new stem-cell treatments in China. Fucking *China*! Almost every bit of him would have known that these Internet adverts for stem-cell therapies that lack a shred of evidence prey on the last grains of hope of the rich, the poor, the desperate. *Nothing almost sees miracles but misery.*

I was marooned by the need to preserve him as he was. There were things I wanted to tell the Schwartz Round but couldn't – most pressing of all, the gap between Lewis's awareness and his illness. And I couldn't tell them because it meant too much to me. I wanted more from my grown-up, fatherly sixth-form friend. I wanted Lewis to look his disease dead in the eye, stare it down. I wanted him to tell the truth, however frightening. Believe it or not, it wasn't until I was halfway through my speech that I realised I couldn't bear his denial – couldn't bear it because it disrupted my own; couldn't bear it because it brought to mind my father in different ways (who couldn't have been further from him otherwise). For all of my naivety and selfishness I know there's something apposite at stake: that like fathers and their children, patients do their doctors an immeasurable service by dying well.

The last time I saw Lewis alive he had been admitted to a geriatric ward here, at his own hospital. Geriatric at fifty. Inappropriate placements were becoming a feature; a symptom of new pressures on bed space. He was suffering with another pulmonary infection, could barely draw breath. He had cognitive problems, evident in his anxious confusion. He weighed seven stone. His recently muscular frame had been obliterated in a few months, including those muscles around his mouth, making speech impossible. The ward was populated by elderly, semi-naked men wandering round like Old Testament prophets. On one side of him there was a bed with an eighty-two-year-old who had advanced vascular dementia, on the other a floridly deluded eighty-eight-year-old with a urinary tract infection. Both of them looked full of life compared to him. The staff hadn't given Lewis any means by which to communicate his wish to go home. A simple pen and pad would have made the difference. It's the sort of thing that he would have thought of.

From the ward phone I rang the on-call neurologist. There was no need for Lewis to remain in hospital, he told me, it may prolong things by a few weeks, a month at most. And so with the life-saving nurses staring at us in disbelief – lips pursed, heads shaking – I wheeled him off the ward and into a taxi, hugging his already cold-feeling body for the last time. He died three days later.

Eleven days after that and two months before the Schwartz Round, my father died. I didn't tell them that either. Instead I told them a story about Lewis, not the whole story, but a story nonetheless, because if I told them the truth, as it was, it would be misleading. So, by accident, I turned him into a portrait of someone who influenced me deeply with his own brand of unambiguous heroism, whose death affected me so much that I was unable to finish the eulogy, had to sit down, let my grief speak instead of me. Everyone understood. Nobody understood.

My dad's death certainly changed things because it made me the next in line. But the death of my friend, of someone who was close to me in age, a doppelgänger but also a father of sorts, changed things more dramatically, meant that there was no such line. Though his life was a beacon of sorts for my own, it was the manner of Lewis's dying that I remember most. My boss recommended that I see the Trust counsellor. I made an appointment.

Denial blew through me in different, paradoxical ways. Lewis shared an illness that was the same or similar to that of many of the patients that I see. But unlike them, I knew him – Gene Vincent fan, stylish left-hander, uncompromising advocate for patient welfare – and cared about him before his illness. And then I watched from a distance as he and his wife and his children suffered more than I could imagine for a year and a half. Something had broken, the distance was collapsing, the old forms of denial were no longer available, meaning, I told myself, that I would never be able to armour myself against my patients in the same way again.

Except that later that week I cancelled my first appointment with the counsellor. I didn't show up for the second. A few weeks after that, as another wife described through tears how her husband, diagnosed with a fast-acting and terminal neurodegenerative disease, had in his innocent-sounding, over-excited play – a kissing, blowing, sucking game with their one-year-old – bitten his son so hard that the boy bled, I wouldn't think of Lewis in his suffering, or how alone and unprotected I felt without him – nobody to help digest the unimaginable – because it would have capsized me and looked to her like something other than it was. Instead I took what refuge I could in the distracting mechanics of *care*. I made a note about required risk assessments, another note to alert local child services and safeguarding, another to find respite care for the wife, and a final one to refer them to whoever was the new neurology social worker.

Murray

'So, what are you going to do with the rest of your life?'

I'd never heard that question before, hovering close to cliché and thumping me with its gravity at the same time. Certainly nobody had ever asked me it, nor had I ever put it to myself. It was his opening gambit, the first thing he said the first time we met, on a balmy evening in July, three months after my father died. I couldn't face 'counselling', so I agreed with my line manager to 'clinical supervision' with Dr Murray Simon – the label makes all the difference. I'd had many supervisors in the past, but none like this. Early in my career it had been fortnightly. I was still supposed to have a dedicated hour every month but though Lewis and I listened to each other talk about cases, I hadn't seen anyone formally for years. I felt the consequences more and more: unchallenged, my work staled, followed a template learned long ago that hadn't been updated, with limited, limiting beliefs about what was possible.

To be honest it was more than that; my sense of self was untethering, the Binding Problem had become personal. A patient I had recently been seeing suddenly stopped coming without notice. He was an interesting man, good company, idiosyncratic, certainly more articulate than many of the neurologically devastated men and women I spend my time with. I took it personally. Rather than cancel his weekly appointments I used the time to write the story of what might have driven him from my room, filling it out week after week with details about the different pressures on him, his character, its strengths and weaknesses, its secrets, until I had a journal's worth of notes. Writing down my memories of him, memories I didn't even have, somehow mollified the loss. It also sprang an unacknowledged genie from its bottle. In writing about him I had stumbled on a silent partner – a prisoner even – I recognised from many of the patients I see, who slips out urgent messages-in-a-bottle while the respectable part says all is well. One evening the line manager knocked and opened the door. I was sitting in the room without the light on staring at an empty chair, my notebook open in front of me. She thought she had heard me talking to someone from outside. She had come to tell me, in her blunt, *realistic* manner, that certain more long-term interventions – like this one – would have to be cut because of budgetary pressures. The sessions with my imagined patient would have to finish. I surprised myself with my display of controlled outrage: did she have any idea how catastrophic this might be for his already fragile identity … ?

If I was my doctor and had to formulate this outlandish behaviour I would have pointed to various predisposing, precipitating and perpetuating factors, among them the attrition of working with a patient group of whom the vast majority only deteriorated, and horribly; the variety was in the rate, and it wasn't just the patients any more; death heedless of clinical boundaries, barging in without an appointment. Then the general tone of disenchantment and dissociation in the hospital: estates in disrepair, services at breaking point, the lengthening shadow of cuts; in the most recent financial report the only service with a healthy profit margin was the Trust car park. It may say 'Promoting Clinical Excellence' on the huge banner under which you walk into the Trust's one new building, but the CQC's recent verdict (buried under MBA-speak in a lengthy email to all staff from our preternaturally young CEO who resembles a male Zoella), was 'Clinically Inadequate'. How can you trust something when its self-evaluation is that far off the mark? Like the whole institution has a severe brain injury, like the whole hospital needs intensive care.

Murray was from a different era. Well into his seventies and basically retired, he spent most of the week in the country with Frieda, his girlfriend, practising calligraphy, fishing and qigong. Only people who knew him told me how he'd been dividing his week like this for three decades or more. And people who really knew him spoke of how many years ago the calligraphy had been an emergency spiritual intervention, helping rescue him from a spectacular breakdown when he was already on his fourth marriage with a huge private practice and an international teaching schedule. Since then he'd kept his life as simple as possible, they told me, while remaining on intimate terms with his fragility. 'You'll get on with him'; 'He's perfect for you', is what they said.

'What are you going to do with the rest of your life?' he repeated.

He was asking me simply, like a father might, but there was also something rhetorical in it, as though it wasn't mine to answer, or that it contained an embedded contradiction – like a Zen koan – meaning I couldn't receive it simply. He leaned in a little, flinty eyes fixing me, hinting at cool and warmth at the same time. The room felt like he'd been sitting in it for fifty years, a hermit's cave, every corner under his control. I thought about the featureless, dissociated feel of my clinic rooms, as much an extension of myself as a new, poorly fitted artificial limb. There was a real sense of force in just being there, a place to be myself while he gently X-rayed me; in fact it was disconcertingly intense: I wish I'd brought a case to hide behind, but I was the case.

He left the question hanging to ask about other things.

'Were your mother's family religious? What does it mean to you to have been born on a Scottish island?'

His questions were familiar but his punctuation made them different, or they took sudden unusual turns.

'What do you think of, the moment before you sleep?'

Were they important in themselves or just purposefully oblique, to read me in other ways? And asked with a smiling, forceful quality – I didn't know the two, soft and hard, could sit so closely together – with the implication that nothing could be more serious or critical than right now and what I was about to say, except it wasn't *that* serious. I'd never seen a doctor with this quality before, as though the treatment had begun the moment he opened his mouth. And it didn't *feel* like a performance.

'Why don't you tell me about a dream?'

I couldn't think of one. Or rather I couldn't think of a good one: I didn't wanted to be associated with boring dreams. So I made one up, based on a film I once wrote, but with improvised additions:

There was only one house in the valley. I had woken up that morning and nothing worked, not the television or the cooker or the lights. Four of my fingers were missing. I knew something terrible had happened in the world beyond where we lived. My parents weren't there to help, but I wasn't a child. My daughters were there: I was the parent.

A tense, silent walk leading the girls up towards the valley-head. The geography looks like Snowdonia or the Yorkshire Dales where I often cycle. It's unusually hot. I'm rushing up to the pass to find out what's gone wrong beyond. I'm dragging the sleepy children up the slope but they are distracted by every buzzing thing that comes into view. Where there should be sheep and heather and gorse, there are terraced paddy fields and lobsters. I want the girls to walk faster.

A wide dual carriageway springs up from nowhere. I check for traffic: it's clear on both sides. I cross. When the girls arrive at the road I am already fifty metres up the path on the opposite side. They step off the curb. Out of nowhere a dark car appears in the distance going impossibly fast. I shout but they don't look up, too used to my voice, too used to my shouting. I run back. The car screams towards them with speed.

I cover my eyes. I hold my breath.

When I look again the car is crumpled in the middle of the road, as though it has hit an imaginary steel wall. It is empty; no driver, no passengers. Walking round to the other side I see the girls getting to their feet, their clothes are torn, they are covered in black oil. Cordelia has a graze on her knee, but otherwise they are unharmed by the collision, still engaged in the same carefree conversation they were having as they walked through the field.

I start breathing again. Only I can't.

'OK. Let me see,' said Murray, as though he really meant seeing. 'You are the girls, unhurt despite your worse fears. You are the driverless car. The children are your children and they are your patients. A father hell-bent on dragging kids into adulthood, as though parenthood was a "job and knock", to be completed in a hurry, years ahead of schedule, Meanwhile the world in all its presence gets lost. You are missing four fingers, the fingers you use to count on for the kids: four decades of your life gone, you're halfway through and its already time to leave. The dream is asking you the same question as I am, because it's the only question …'

He looked every inch the oracle in that moment.

'I'm heading for a wreck – it's predictive.'

'You're compelling yourself towards it. You create the conditions for your own crash because you believe it might save you. There are signs of a different sort of life after the apocalypse; the TV might have stopped working but the valley itself is thriving.'

'And the kids?'

'They want to play in the filth. If you have to leave then leave, it's what fathers do, then they come back.'

'Leave the girls?'

'Part of you has already left. You might have to go in order to come back properly. The children have survived you, they can survive without you.'

I motioned to get up; after thousands of appointments I had an improbably accurate sense of how long the clinical hour was.

'One more thing: don't make things up any more. The lobsters ... ? Please. When a story lies too much it strangles the life that supports it. Or at least lie better, lie therapeutically.'

How did he know?

'Go find somewhere and shut up for a while, let silence do its silent work.'

The sense of having ground beneath me again, someone to be contained by, the beginning of a conversation that was really a form of permission at a deep level. I remember cycling home over the Albert Bridge at dusk, the hanging baskets of red geraniums darkening visibly next to the luminance of the neighbouring irises – a compensatory spectrum-shift known as the Purkinje effect – and with that the thought that it is only once in every thousand nights that I notice something of the wonders I know, not counting the much vaster universe of things I don't, that won't ever be noticed. Not having picked up a drink in years, I felt drunk on his powers; the intuition, the empathy, the wisdom. It was as though he'd lived my life and then grown up afterwards. He saw himself in me, I was sure of it. By holding me in his mind, by holding himself still, he was showing me what I might do with the rest of my life. I found myself pedalling quicker, wanting to get back and write everything down. It hadn't seemed right to do so at the time, especially when he took no notes. As soon as I got home I wrote everything I could remember. Then I lost what I'd written.

I felt a little uncomfortable when I saw him the following week on a typically grey, viral-feeling summer afternoon. I didn't know how things could move on from our last session, it was complete in its own right. They say you should never go back to a holiday romance, and it felt a little like that.

Things started generally. I told him about a difficulty I had coping with the angular manner of one of the clinical fellows. My heart wasn't in it. Or rather I sensed that his heart wasn't, like he'd gone missing. There was a long, freight-less pause.

'Were your mother's family religious?'

A knot in my stomach, a child that hadn't been heard, hadn't registered in his caregiver's mind. I tried to let it go, but the anxiety meant I was falling inwards now, our actual conversation somewhere far away, beyond reach. Later in the session he asked which part of England I came from. Or worse yet, a caregiver's mind that couldn't register. *He* should have taken notes.

Still ruminating two days later I called him from the hospital at his home in Suffolk. Frieda answered. He'd gone to the local shop for some milk. I explained who I was and what I did.

'Can I ask you about Murray? I hope I'm wrong about this.'

★

This time the room was mine and that changed everything. He was older and smaller than two weeks before, eyes dull, no longer transfixing. I had the familiar feeling of a patient slowly disappearing and my powerlessness to do anything more than recognise it. It had been a performance after all: his curtain call, his valediction.

The theory of cognitive reserve has gathered traction over the last decade. It claims that the more 'high-functioning' the patient then the greater the buttress against neurodegenerative change. That's the theory that has spawned a geriatric Literature on the beneficial effects of crosswords, sudoku, knitting, jigsaw puzzles, computer games, etc. To extend the life of the mind? Or a permission for the regression we long for as we approach the end? Either way I'm distrustful of theories that fit too closely with what we want to believe: that ageing changes nothing. But with Murray sitting in front of me there was one study in the Literature that came to mind: comparing reserve across different jobs it found that musical conducting was the profession with the highest associated reserve, the hypothesis being that conducting relied on 'analytical', 'emotional' and 'physical' networks communicating via the association cortices, different faculties combining which is somehow preservative: conductors' brains are their own symphonies. Hadn't Murray lived different lives drawing on a diverse range of faculties and skills? He should have stellar levels of reserve.

His verbal intelligence was predictably off the scale, and his perceptual reasoning was unusually high for someone as literate as he was. Things looked a little shaky on his visual construction; there was something careless about his copy of a complex geometric figure which could easily be temperamental. But, as I knew, he couldn't hold on to much of a story I read him. I read it again, I wasn't supposed to, but I was hoping I was wrong. He got the first few words, then fell silent, then he started making it up, confabulating, just like I had with the dream (I can't stop seeing myself, my father, or their shadows in every man I meet). Murray discerned a story where there are only disconnected words, while other patients give back as unrelated words what belongs together as a meaningful whole: the brain fragmenting in different ways towards the same end, 300 different dementias and counting, all eventually fading to the same grey.

Afterwards, as we joined Frieda in the waiting room, he wanted to know how I was getting on, asked me about the girls, told me how he'd been invited to another conference on the back of a recent talk he gave on the I Ching.

'That's really great Murray, they're lucky to have you.'

'Will I see you at my place at the usual time next Monday?' he asked.

He'd been docile during testing, following my commands without question, and now, poignantly, he was trying to reclaim some territory after all the power had been ceded. It was cruel but the force had left him and I was leaving him too.

'Or I can do Tuesday if it's better?' he offered.

Suddenly, he was an impostor of sorts. Our doctors are nothing without us.

'Thank you, Dr Benjamin. Come and visit us in Suffolk sometime, bring the kids,' said Frieda, at his side, an instant adept at the sort of deflection her partner's disease would require from this point on.

The payoff for the reserve hypothesis was that in exchange for the buffering, when decline showed in those with high reserve, it came quickly and dramatically. The disease had been at work, the pathogen likely germinating in his mid-forties, my age now, with various as yet undetectable insidious biomarkers working hard behind the curtains over the next two and a half decades, before Frieda thinks that she notices him repeating himself more than usual, and long before he will notice anything himself. And so that by the time the disease has poked through, it has already done most of the damage unseen, devastating the reserve, his functioning on the edge. The building was almost empty; the last thing to leave was his pride. I had lied to him about the dream of my future, but now I was telling him the truth about his – diagnosing him – shining a torch on a short pain-filled path which he will walk without me. By meeting him I had killed him – and eliminated another support, a possible guide for me in the process – spotting what might have slipped past unnoticed or accommodated at least for a while, transforming him with my contaminating sight from an artist of sorts to a foolish fond old man in the space between sessions. That was what was left of the rest of his life: he sensed it I now believed, when he introduced himself by asking me that question. And what would I do now with the rest of mine, without him?

Ben

August. My room is ten degrees hotter than it is outside, as hot as it is cold in winter. The hospital is half empty, the consultants on mass holiday. Nobody gets ill during August, people are either alive or dead. Or that's how it looks. Really summer is the season of head injuries: helmet-less cycling, drunk cycling, drunk driving, stage-diving, drunk swimming, very drunk sex, pub-abseiling, pub-bunjeeing, drunk base-jumping, drunk free-basing, base sex-jumping, freebase drunk-jumping. Then the sub-specialty of drunk fighting; a ten-cubic-centimetre donut with jam inside, for punching, kicking, nutting, bottling, bricking, baseball-batting, or if the fight turns domestic then microwaves, irons, kettles, NutriBullets – an Ideal Home Exhibition of head trauma. All together it makes August a celebration of smashed cerebra, a festival of fractured phrenia: WHAM! … Club Tropicana neurology. It's as though the sunshine makes people long to drive round in open-top heads; decompressive craniotomies makes convertibles of them, leaving deep pink tattoo-like scars – a crescent moon, an anchor, a ham end, a misshapen rainbow – and a lucky dip of assorted cognitive and motor impairments. Unless you are Polish or Albanian or Magyar, in which case it is always head-injury season: January, February, March, April, May, June, July, August, September, October, November, December ... But try doing it backwards, one month after you've fallen twenty feet from scaffolding on a litre of vodka – you won't get out of autumn I promise, nobody does: it's always October for them, September at best, Christmas if they're unlucky – even in August.

He looks frightened, this middle-aged man, just a few feet away now, across the wasteland of my desk – same age, similar clothes – just the other side, the wrong side. He's been here before, many times, his file fat as a telephone directory. Sometimes there is no head injury, no stroke, no tumour and still the person's functioning can be every bit as compromised as the neurologically devastated. Only I wouldn't think of putting this wretch in an MRI and expect it to tell me anything, expect to see a brain materially different from my own.

It's just so brain-poachingly hot in here. While he searches in his pocket for something I send an email to Estates letting them know it's eighteen degrees above UNITE-specified thresholds. There is an automated reply; the manager will return from leave on 1 September ... If there's anything left to return to ...

What is his face asking me? On the surface it is cold-looking, mask-like, Parkinsonian, but something far beneath thrashes round like a severed electric cable. Another one wanting to be discovered, understood, told and then retold ... but I'm not in telling mood.

'FUCKIT ... *CHICKEN* ... FUCKIT ... *CHICKEN* ... FUCKIT ... *YOU FUCKIT*.'

The muffled sound of the Tourette's clinic from next door. A couple in their twenties, I saw them in the waiting room when I was collecting him, a diagnostic double act.

He wants to take me back to a night more than twenty-five years ago, when he was nineteen, which he believes is the apex episode in the story of his life: expressing so much of what had gone wrong before, shadowing everything that came after. The 'attempt' led to a week-long stay on an acute psychiatric observation unit where every session was taped, a caveat in the confidentiality agreement: in part risk management – admissions were much more likely to suicide than outpatients (the paradox of 'care'); in part for training purposes. His litigiously minded law student friends would insist, several years after his discharge, he be given a copy of his own. He has it with him now.

'Can we listen to it?' he asks.

'CAMEL … *BOLLOCKS* … CAMEL … *BOLLOCKS* …' A lover's discourse through the thin wall.

His tape, his precious tape, a message sent from how bad things can get to now, like bottled 'Essence of Despair' for a fragile future self who might be eased a little by its scent: as though today's 'bad' will seem less so next to this recorded 'worse'. He places it in an old cassette player.

A voice starts up, his voice from a quarter of a century before:

'Words, words, words,' that's what Dad always says when I try and explain how my head is, 'Words, words, words,' as though there are other options … He throws this faraway look, smiling – always fucking smiling – makes his eyes shine like Emperor Ming … Nobody has ever known anybody's face as well as I know his: the tells, the early warning systems.

God he looks tired, beaten, his overgrown beard silvering in patches. I imagine the father listening to him; the thin, twitchy voice, hard to hear – less plodding, less medicated-sounding than now – that empties his words, makes them riderless.

He likes drama. I directed *King Lear* last term. He came to see it three times in a fortnight, smiled all the way through like it was a sitcom … He loved it, of course he would. 'What about me? What do you think of me?' My father,

Lear Junior ... I cast myself as Edgar; there's a few of us going mental in the cloister gardens for the Japanese tourists, but I worried the cast were too Brideshead, they needed a rougher, madder example setting ...

He can't stop talking. The traumatised child learns that language is always ineffective when it comes to expressing internal states with any precision, that life is so hopelessly fragmented it might never be shaped and smoothed into a single narrative, with effects neatly dovetailing causes. It's much messier than that; what holds true one moment will almost certainly be illegible the next, forcing endlessly prolix approximations, exaggerations, fabrications even. And what other option is there? The silence of Cordelia? It's an attractive possibility: if I say nothing I might create the impression I understand everything.

'COCKFOSTERS ... *FUCKIT* ... COCKFOSTERS ... *FUCKIT* ... COCKFOSTERS ... *LAGER* ...'

I threw up on two nights as Poor Tom, wet myself, dislocated a shoulder, screamed so much one eye bled inside ...
I accidentally stuck my sword Excalibur-style into Edmund's head in our fight scene, he had to go to hospital, twice ...
Words, words, words? There *is* another option. I go into the bathroom, grab a full bottle of pills, any bottle and empty it into my mouth. Rock and fucking roll.

He is staring out of the clinic window at the wretched Portakabins, we both are: the 'new' home of the Renal Unit, inundated with nutgrass and ferns, like the hovel on Poor Tom's heath.

Except after I've thrown it away, I pick it out of the bin, '60 Paracetamol' – for future reference – then I go back into Dad's bedroom.

'Future reference'? ... Clinical sensitivity means they don't let the camera crews anywhere near the suicide ward. But for some that doesn't cancel the show. These patients have CCTV permanently rolling in the corners of their minds, because really there is nowhere left where the cameras can't go; the audience self-generated because there has never been a parent to watch or hear them.

Are you listening?

A long pause on the tape. Then:

Can you hear me?

For a moment the twenty-five years between the recording and now collapse and I think he's addressing me.
'FUCKTAGIOUS ... *CUNTATHON* ... FUCKATHON ... *CUNTAGIOUS* ...'

I told him what I'd swallowed. I told him I wanted him to concentrate now, because I wasn't going to do anything about it until he proved that he could hear me.

I think of Edgar channelling Poor Tom, the madman, extending the performance to every corner of his psychic life.
The tape has warped with age. I think the voice is saying that as the pills metabolise in his stomach he sets his father a kind of quiz: 'on me' – the questions requiring him to remember early birthday presents, favourite crisps, schoolfriend names, pets and their burials, as a kind of paternity test – which I presume, like taking the pills, is really a way of proving his own existence in this zero-gravity moment and which the father has to complete before he will allow himself to be driven to hospital ...
I've had enough of this hackneyed drama, another dialogue which is really an endless, endlessly self-involved histrionic monologue. I see very clearly what is happening here, that he is taking me hostage just as his younger self kidnapped his

father, meaning there were four tired, angry men in the room now, wrestling each other for anger rights.

'FUCKIT ... *YOU FUCKIT* ... FUCKIT ... *WE FUCKED IT* ...' A fifth and sixth if you include next door.

Why did we never go to a football match together?

Suicide can be so innocent an impulse: the paradoxical, life-preserving belief that one is trying to remove that part of oneself that is so wounding, forgetting about the baby in the bathwater.

'The operation on my leg? Remember? Just get one right, Dad, prove that you ever paid attention.' He's not smiling now. He's trapped, unable to remember anything about me, and therefore I'm trapped too. Then he starts crying; huge, dry, tearless heavings.

I am thinking about the end of the clinic, of being in the cool of the lido for an hour, underwater, of silence. The week before, an official complaint was made: a patient alleged that I was reading about Arsenal's summer transfers on my phone while he completed a potentially life-changing matrix reasoning task ... He waited until after the clinic to lodge his complaint. Turned out the patient was a doctor in disguise.

It was a second major strike, the first nearly two decades earlier, when I had slapped a violent service-user in self-defence. Like our patients, our medical histories won't disappear. I was waiting to hear about my punishment.

I try to stop listening, but I can't keep his voice out.

It's freezing outside. He's still in his pyjamas. We are sheepish around each other now, both of us having spilled our guts out − nearly. It's only a short drive to the hospital; we've been loads of times before. The distance between us is

opening again. Listening to the Cortina's old engine splutter –
it won't start in the cold – we both hear the jokes we might
have made but don't.

'I'll have to ask Ian,' he tells me. Ian lives round the
corner. He always stays up late watching repeats of the racing.
'No, not Ian,' I say. I tell him I'd rather run to A&E ... I'm
a terrorist in my own house but a total coward as soon as
I'm in public. 'We can't ask Ian,' I say.

Because his voice is close to home: I know these people,
people like them.

We sit in silence for a moment then he gets out of the car.
'Promise you won't tell him?' I shout after him. Five minutes
later we are in Ian's car. I know that he knows from the
way he's looking at me through his rear-view mirror. I see
his contempt. And then I think of Lee: Ian's son, four years
older than me, bouncer's leather jacket, two earrings, third-
hand white Capri, no O levels, weight trainer, pussy magnet,
apple of Ian's eye, Prince Lee from Leeds – died sixteen
months ago of a bleed on his brain. I see his contempt and
I feel it too: I don't blame him.

It's the same look I see in the middle-aged nurse at the
hospital as she jams a sharp-edged plastic tube into my
mouth, like what I've really swallowed is judgements not
pills, swallowed and digested.

I think how I must be looking at him in this way too,
twenty-five years later. Nothing changes.

'I can't fucking breathe,' is what I tell her, only it comes
out like 'Aggaggggaaaaakkiiinnnbweeeeeeevvv' because the
tube is filling my mouth and when it's finally scraped its
way into my guts she asks me if I know what paracetamol
does to someone's liver. If I've ever been on intensive care.

If I know how much it costs to look after someone like me for one night ... As though I'm the one doing a quiz. And I want to tell her what an evil witch she is, that if she ever has a child I hope it's made of spleen, that she can take her beloved NHS and ... but I can't speak because of the tube, I have to shut up.

And somehow this is what I need, to stop talking ... and a receptive audience to hear me stop: someone to tell me what I've done is wrong, and someone to look after me at the same time ... All of a sudden I'm nine years old and back in my bedroom and.

The voice stops abruptly: 'and'.

Silence, recorded silence, compelling but restless, as though a bonus track might suddenly jump out of nowhere.

'ONE, TWO, COUNT MY TITS ...' A different solo voice from the clinic next door, a male voice.

It's dusk outside. He is still lost somewhere out there. Past the Portakabins the rest of the hospital laid out like a patient awaiting resus, anatomically dislocated, as though reconfigured by a drunk, maniacal surgeon under the instruction of a dysmorphic management consultant: breasts inside, kidneys out, Bloods leaking, Infectious Diseases everywhere (including those – MRSA, E. coli – of our own making), Lungs collapsed, asbestos found in the lagging, Liver open to the air, Eyes half closed, Ear, Nose and Throat swapped places, Cardiac Care non-existent. Psychiatry a teeming hovel on a blasted wasteland miles away from anything. Only the Brain was new, new-rology, towering above everything: invested-in, expensive, decapitated, a giant totem of Cartesian priority, of human uniqueness, crowned with helicopters delivering us from bad dreams (a rail fitted around the helipad perimeter to discourage potential jumpers), opened in a hurricane by a celebrity former patient, re-enacting an emergency, his smirking face over-topping red stretcher blankets.

The tape recorder – his time machine – clicks off. He takes out a new cassette from his pocket, and replaces the old one. This time he presses 'record'.

'Do you mind? My memory is shot.'

He starts to speak now, going back to the late 1970s, closer to origins, to possible causes.

First there was just the name to go on, he explains, but what a name. Later he heard the voice, the hoax one on that tape, the one sent to the assistant chief constable that started in a thick Geordie accent 'I'm Jack. I see you are still having no luck catching me.' – 'Remember?' he is asking me.

'ONE TWO THREE, COUNT MY DICKS ...' from the clinic next door.

I know this territory well. For a certain generation of middle-aged men, Yorkshire Ripper fantasies are almost a diagnsotic idiom in their own right; delusions which emerge from the foggy fabric of the real (Bollas 1995).

He recalls how he and his friends would spend lunchtimes in the playground perfecting their impersonations. He could have told the police it was a fake, it didn't sound like *him* one bit ... Did I remember his voice, he asks me? I thought I could, but it verged on neuropsychological fiction; we were both too young to remember most of this, there can be no memory without a well-developed grammar to support it. So if it wasn't our memory, what was it? Whose was it?

Leeds, his home, was not the epicentre, he continues, but by 1978 the Ripper's ambit was as large as West Yorkshire. *Fourteen and counting.* The Black Panther was up the road in Bradford. And Saddleworth Moor was a short trip over the Pennines on the way to his Auntie Bev's. Beyond the bravado of the playground, it felt like he was closing in: at home he wouldn't go up the stairs on his own unless all the lights were on. He stole a torch from the corner shop. Then the police released a photofit. It was terrible as a likeness, but still terrifying. To the extent that it could look like anyone it looked a bit like him. He started to sleep with a small knife under his pillow.

My anger has dissipated. I am spellbound by the mix of horror and nostalgia. But it was more than that: the memories of my father were disappearing so quickly that this man, or rather this shadowy, hologrammatic figure takes up occupancy, parasitising on what remains of Dad, becoming more and more real as the moments pass. Meaning the distance between myself and the patient is closing, that we are on the same side of the desk now.

His parents never argued openly; so much of it was just felt in the heaviness that clung to every inch of the smoke-blackened terraced house on the Stanningley Road. But the cadences of arguments, of recrimination and despair, could sometimes be heard behind closed doors, through thin walls – distorted, threatening – like the nine o'clock news headlines, 'Idi Amin', 'The Khmer Rouge' the first cases of 'AIDS', which he half heard from his bedroom. This was how he found out that his father had been stopped and questioned by the police, twice.

Whereas his parents' bedroom was empty apart from an old copy of *Linda Lovelace*, long unopened at the bottom of a drawer, his friends' mums and dads' bedrooms were erotic cornucopias: porn mags, dildos, lubricants, *kit*, only half stashed. This was the secret of a good marriage, of parenting; openness, imagination, fullness, pleasure – carelessness within limits; the opposite of their anxious, secretive, empty protectionism. It didn't work either; there was no protection. The world was broadcasting its aggression; new video films that were banned because of beheadings, drill-impalings, snooker-balls-in-sockings, his friends' parents had them all on VHS (while his watched *Jesus of Nazareth* on Betamax) and they'd slo-mo gory bits over and over on play-dates, only they weren't called 'play-dates' then. Not one of them got inside his head like the Ripper had. He was a surgeon too, with his hammer, his hatchet and his gloves. All the talk was Global Armageddon but he wasn't bothered about the end of the world. He was in his own arms race: now he had a Bic razor, a screwdriver, a carving knife under his pillow. One morning he woke to find dried blood on his hands: he'd cut himself while he was asleep. He'd tell his mother he'd done it at school.

His father had steadily put on weight over the last few years, undiagnosed depression most likely, all those corpses weighing him down. He would sit in front of TV sport for hours eating wheels of cheese, whole loaves of unsliced bread, like Gulliver, the crumbs nesting in his thick black beard, like Fred West who would eat whole raw onions to douse his appetites. He made numerous attempts at developing a jogging habit, which was largely frowned upon back then in that part of the world, but serial killing was a young man's game. And he needed to keep up with his wife, who was following aerobics classes with someone called Mad Lizzie on the television – another psychopath presumably – and had started a night class at the university. They both knew the end was coming, so best get in shape before they started putting themselves about a bit. It meant that neither of them were really home, even when they were.

His father was a dreadfully early riser, still the middle of the night really, his jog was usually well over before anyone else was up. One night, scuttling to the loo without his torch, he saw him through the gap in the door coming up the stairs, drenched in sweat and blood, desperate-looking, gasping for air like a big fat fish.

He pauses the tape; to catch his breath? Sweat is pouring from both of us. It's momentarily shocking to think that we, fathers ourselves now, are older than our fathers were then.

'CRABSTICKS? CRABSTICKS? PLEEEEEEASE ... ?' next door's patient pleads with his doctor.

I think how often people with these kinds of memory disorientations appear less obviously abnormal than they actually are – hold down a job, a relationship, at least for a while. And as I am thinking it, the thought slowly protrudes through some thick northern fog (Brodmann area 13?) that I have seen him before – the outstanding forehead, the dulled eyes, the tombstone teeth.

He starts again, remembering the night when Jacqueline Hill was murdered in the car park behind the shops on Otley Road, how his mother had been in the same car park half an hour before; they were both students. His dad was out that night as well. They probably had an argument earlier that evening, like most evenings. The two women resembled one another. Had his father made a mistake? He tried to keep his brother James awake as long as possible in the bunk below. They'd run through capital cities first, then the most scary-looking football players (usually Iranians) in their World Cup album, who was the hardest boy in the year? The prettiest girl? What you'd do to her? James would always fall asleep first, leaving him to answer his own questions.

He'd still be awake when his mother went to bed, feeling like he'd never fall asleep again, and then half an hour later he'd hear his dad's heavy step on the stairs, hear him pause at the top, the boys' bedroom to the right, his own to the left, catching his breath ...

The first time must have been around then, around the time of Hill's death, his last murder. For some reason, guilt no doubt, he bought him a train set.

A train set?

'TA-DAH! FUCK-JOB.'

That Ben ... ? Junction Box? The boy who shagged train sets, the boy I never met? Which means *that* father, the one who wouldn't stop smiling?

The door opened and he came in, still breathing heavily. His hand felt for the largest knife under the pillow. He scrunched his eyes shut and stopped his breathing, like he'd practised for years. That would always be his strategy in the event of murder, to pretend that he was already dead. He felt him standing at the foot of his bed.

He didn't abuse him? Was that there in the smile and I missed it?

He imagined his father wanted to confess, tell him everything that lay on his heart. Next door she didn't understand, none of the women would ever really understand, and there were lots of them, so it was his job, a son's job, to know how it felt to be him: whatever else he was or might be, he would always be his father ...

'I want to open my eyes, let go of the knife, put my arms around him, let him know that it was all going to be alright, I'd forgive him, whatever he'd done ...'

No he didn't abuse him. Not in any conventional way ...

'But I can't speak, the one moment that really matters ... So from then on all I'll ever have is empty words; words, words, words, and nobody to hear them ...'

The Ripper still alive, your father dead for what feels like years, estranged again from your mother. The memory of a serial murderer a way of explaining how murderously empty you can feel, another way of giving the numbness of your childhood an electric shock or a whole bottle of paracetamol, inventing fantasies of patricide which might be no more than resisting the impulse – an impulse he put in you – to save him.

Really, Ben is not so different from the rest of us. He has cleaned himself up in every way he can, stopped drinking, drug-taking, eating dairy, started cooking classes and meditation, joined a book club. But still he feels like crying in front of his children without knowing why, lives in persistent, half-realised fear of affecting them in inescapable ways, feels how his job has become a dull stone; a serial monogamist – somehow allergic to the intimacy he craves – getting through people with increasing speed, in fear he might hurt them; has tried every therapy imaginable, they always fade after hopeful beginnings; an insomniac, a binge-eater, a hypochondriac, an illness-denier, addicted to a cardio programme requiring him to run through Hyde Park in the middle of the night with the dozens of other maniacs who are out there too; strangled by the past, losing whatever murky grip he has on the present because he is tortured by the only future he can imagine ... Never mind the rest of us, he is like me – if on a good day I am honest about how bad the days can be.

He comes here for help but really all hope of transformation has long gone – I am nothing but disembodied ears to him, if I am there at all – because whatever he does he cannot escape the feeling that all the possibilities of his life have already been circumscribed by a father, who is always killer and victim – the two as fused as doctor and patient – a dead father who, people tell him, he resembles more and more each day. A father himself, but like his father, always really a son (as long as he had been alive at least there was the *permission* to remain a child) who has inherited an acute sense of being a fraud, running through him like a fault line, as stark as a neurological fact. Which is why he made his tape, a self-portrait in a convex mirror: he needs someone to recognise him, to catch him, to stop him.

It's late in the day, some people can't be helped ...

Craig

… or help is not what they're after:

'I need a psychiatrist.'

'Let's talk this through.'

'Now.'

'…'

'Mental health? Get serious; this place is an abattoir. I want to see the psychiatrist. I want an MRI.'

'I'll speak to the consultant as soon as we've finished talking.'

'I'm not talking. Call him now.'

'He's not on the premises.'

'I want the fucking top man.'

I'm reminded of him because even though it was twenty years earlier he felt like Ben, only Ben turned inside out, shouting everything he couldn't say.

'I want one of those new MRI scans.'

'We don't have an MRI.'

Or a CT scan, or an X-ray, or security cameras or a panic button. He is drawing furiously on clinic paper. He looks insane. He has a badly shaved skinhead – bald patches, clumps of longer hair, open sores — earrings, tattoos. He is wearing a red Mylar suit, the type you might wear if the world was really ending.

'I'm sorry about what happened,' I say.

'I tell her I want to cut my head off by three o'clock every afternoon and she forgets our appointment. If I cut *her* head off she wouldn't forget. This place is a slaughterhouse.'

The walk-in addiction service was overrun and badly under-resourced; some things don't change, or they get worse. I'd started training not long before. We only got one session of psychiatry a fortnight which is why this twenty-nine-year-old man, my age at the time, in the grip of an acute episode, was allocated an assistant counsellor. She had forgotten his appointment. These things happen.

'Didn't I score high enough on your risk thing?'

'I can understand you're angry.'

'Oh you can *understand* can you? What a talent. Nod, nod, nod: born to nod. My head is going to fuck up your MRI.'

Anger and exaggeration: the exchange rate with truth forever broken. His anger felt old and young at the same time, like a nine-year-old in an adult body, younger than that even. He is still living with his mother.

'Have you been using today?' My pen hovering over the tick-boxes on another risk pro forma.

'I am going to use *you*, moon-face. Get me a doctor.'

He is busy with his drawing. Anger in the child as a way of wresting back predictability from a parent who is wildly inconsistent in her love, as a means of guaranteeing rejection or withdrawal, because things are so desperate that any form of guarantee will do.

'You have never seen a brain like this before.'

What if anger were actually a means to stop you feeling something? A fear of disintegration, of going mad; as though sensitivity could attack itself and in that way become less sensitive (after Eigen 2004).

'Come on, put me in your big metal coffin.'

He wants an MRI; well, join the queue. It's more than likely that a wide variety of difficult-to-explain conditions are anchored in aspects of early parent–child interaction – addiction, post-concussion syndrome, irritable bowel, ADHD, chronic fatigue, non-epileptic attack, personality disorders – and yet increasingly it is neurological and genetic explanations that are most sought. We are not enough. We want our brains to save us from this lack. So, madness because sometimes a diagnosis and medication are what people want more than anything in the world, including being understood. Like a blow-up lifesize mother doll: however unreal, however inflationary and unsatisfying they may be, they are still a way of getting held.

He pushes the notepaper over to me.

'Colour *that* in,' he tells me.

A poorly drawn, inordinately complex Venn diagram, like a mutated Olympic symbol. Larger circles overlapping, within them more circles which overlap with themselves and the larger circles. Sprinkled over them are numbers, from 1 to 35. On the side a key with titles for each number.

In the centre are several interleaving Axis 1 disorders: clinical depression, generalised anxiety disorder, panic disorder, agoraphobia, obsessive-compulsive disorder. Around these are a nexus of Axis II disorders: borderline personality disorder (with a question mark next to it), histrionic personality disorder and narcissistic personality disorder. Then there's adult ADHD which is circumscribed by bulimia, overlapping with various substance-abuse disorders – alcohol, cocaine, MDMA, ketamine, ayahuasca, amphetamine, skunk, opium, glue. There is a ghostly circle nearby: seizures (alcoholic? epileptic?), Korsakoff's, alcohol-related dementia, cognitive changes associated with class A drugs – with three question marks. On the other side there are mild-moderate traumatic brain injury and post-concussion syndrome, which both link with anxiety and depression and, via a dotted line, to early-onset dementia and seizures and alcohol. Then psychosis: 'First Episode – 10 years ago, Second Episode – RIGHT NOW???' In a corner there are a whole series of question marks, titles without specifications: 'Triggers? Stressors? Gene structures? Prenatal, perinatal trauma/intoxicants? Early abuse (sexual? Physical/neglect?)'; around all the circles a meta circle, keyed as acute attachment dis-regulation disorder, something I hadn't even heard of.

And there in the middle – point omega, the epicentre, the blank blind spot contained by all the circles – there is no number, no label, just an initial 'C' for his name, Craig (though he doesn't feel like a 'Craig' to me): incidence of 1, prevalence of 1, morbidity unknown, a one-man Gordian knot of diagnostic complexity, a live formulation in five dimensions, the Nobel Prize of pathology, his own chapter in the *Diagnostic and Statistical Manual* (*DSM*), the collective frontier of psychology, psychiatry, biology, neuroscience, psychoanalysis, ethology. Soon a Brodmann area will bear his name, then a syndrome, a street, a first-day cover stamp with his brain in profile: 'Craig's disease', the Mount Olympus of insanity; a condition of pure self-reference.

Despite his singular claims and extraordinary rage, he looked familiar enough to me, like I'd seen him on countless occasions before. This exceptionalism was almost a condition of being a patient. Each of us contains the entire *DSM* but we must somehow make our illnesses special, our own that is, otherwise they remain terrifyingly separate from us. Viewed in this way illness is just another product to express who we are, like a car or a haircut or a dance move, signs of our freedom, triumphs of individualism. The trouble is that we quickly run out of products – the *DSM* shopping catalogue lacks variety – forcing us to invent new haircuts to keep the sense of ourselves afloat, give ourselves names that are ours alone. A doctor friend of mine had a daughter in a large class in a Hackney primary school. The last girl on the register was called Unique. One day a new girl joined the class, also called

Unique: Unique II. Likewise when I contact HR about a payslip error I find out there are thirteen other 'A. Benjamins' on the Trust payroll. So 'Craig's disease' ('Craig'? Seriously … ? The most common of all proper names) first entered in *DSM* (sixth edition 2021) immediately after 'Benjamin's syndrome', denoting 'an obsession with the singularity of your diagnosis while fearing that any specific diagnosis is too narrow'. In other words, madness's way of having its cake and eating it – terminal uniqueness.

The scientific method provides us with a parsimonious antidote to bullshit-complexity and rationalisation:

'Well, Craig, this is an addiction service. Let's call your complex-looking diagram Hypothesis A, and "Common or garden drunk" Hypothesis B. Now, as I don't understand A, let's run with B for a while.'

Is what I want to say. Instead I work my way methodically through the diagnostic boxes on the pro forma, ticking practically all of them, just as he demanded. In this way I ensure the department is paid for the service I've provided and I create an indelible medical history which in the future, only doctors, policemen, insurers and potential employers may have access to.

'I'm going to need a major blowout after this …'

That's right, rationalise your future behaviour, addict yourself to whatever causes you most pain, as though the addiction includes the will to exhaust it. But exhausting this much pain will take more than one lifetime.

'I want fucking sectioning or …'

As he speaks I notice that my heart is strangely at its resting rate, my mouth is not dry, I am not sweating. I can see and hear everything clearly as he continues to shout and swear his head off. The Literature indicates that male swearing decreased from 1,000 words per million in 2008, to 500 in 2018; by contrast women's is on the increase. It was as though he was trying to give me something of what it was like to be him, to place a piece of himself in me, only I rejected the offering.

I imagine punching him in the jaw mid-sentence, as a way of shutting him up. I don't want to, it's just what comes to mind. I imagine him gulping in a huge lungful of air in shock and then bursting into tears. I just imagine how that would feel. Then I look around to see if anyone is watching. I look around once more. I imagine it. I think I just imagine it. Then I leave the room.

It will be his word against mine.

Brad76

'Any major illness?'
    'No.'
    'What about in your family?'
    'We live well into our eighties.'
    'Both sides?'
    'Yes. You're funny.'
    'Any learning difficulties?'
    'No.'
    'Motor difficulties?'
    'Stop.'
    'No … ? What about music?'
    'All sorts,' she replies.
    'Can you be more specific?' I ask.
    'Is it important?'
    'It may be.'
Her hands move in a way that isn't connected with what she is saying. The photograph did not look like this.
    'I feel I'm about to say something wrong,' says Woman 3.
    'There is no right or wrong,' I tell her.
    There is.

★

If it was madness it was mine, but it belonged to us all.

The sky was too blue to be convincing. I looked at them lined up in silence, 200 or more, by the lake's edge on this late September morning in North Wales. The children would arrive soon with my mother. We'd had 'Team Benjamin' T-shirts printed: it was a joke and it wasn't. It was cold, as cold as it was hot in London, meaning serious kit: five-milli-metre wetsuit, neoprene balaclava, gloves, boots, goggles, chafe cream. Seen through the dawn mist which hovered above the lake they looked like a low-budget alien aquatic army ready to take on Flash Gordon.

The Brutal: '*The Toughest Iron Man in the World ... A lethal 3.8k swim in a freezing slate tarn; a kick-ass, hill-stacked 180k bike ride including four separate circuits over the treacherous Llanberis Pass ...*'

Something pitiful, childish, bathetic, about these infla-tionary claims. They worked in the opposite direction, made the world seem like a country fete. And *so* anachronistic: 'Iron Man', like a freak-show turn, alongside Bearded Lady and Eel Boy.

'*... finishing with the small matter of a killer 30k hill run followed by an* insane *12k night-time ascent of Snowdon – the highest peak in England and Wales.*'

'Insane'? Do they mean it? Really? It was tiring enough reading the IM Literature; the authors, hands bleeding with clichés, as they attempt to extract awe, iron awe, from their readers, so many six-year-olds tugging their fathers' arms.

How did I end up here? It wasn't supposed to be like this. Like everyone, I'd taken up exercise as a way of stopping my body from doubling or decrepitating, and more pressingly, as a way of managing my mental health. But a light jog here, a hatha class there was never going to be sufficient. The last few years had been difficult; my tolerance had increased, my sensi-tivity gone the other way, meaning the required dosage had escalated beyond control, so that one morning, this morning, I wake to find myself knee-deep in a freezing lake about to

brutalise myself in unimaginable ways. ('Brutalise'? 'Unimaginable'? Really? I'd caught the inflationary virus.)

They looked like Gormley iron men staring across the
water in different, counterpointed directions.

Nobody said a word.

Who are they, these men and women turning recreation
and play into another form of suffering? Did I want to be one
of them? Did I want to be made of iron? Cold? Hard? Insens-
ate? Was that really a long-term solution?

Nobody would make eye contact.

And why was I asking all these low-level rhetorical questions? Anti-rhetorical even; the answers bland, obvious, un-
important. It was a new symptom. Something about my
condition rendered me super-sensitive to motivation. In the
normal course of things life was set up to make self-reflection
impossible, it was only the patients who got scrutinised. But
here, now, there were no patients left to hide behind. Either
that or we were all patients.

'How old are you?' I ask.

    'Wow. Direct. OK. Thirty-seven, nearly thirty-eight.'

    'Is English your first language?'

    'Ha! Just about.'

    'You seem nervous.'

    'Yes, a bit. Are you?'

    'Would you like a glass of water?'

    'I'd prefer something a bit stronger.'

    'Now? Here ... ?'

    'Why not here?'

    'How many units do you drink per week?'

    'Wow, get you. Not sure.'

    'Just an average, please.'

    '... What? Are you writing this down?'

'No ... Yes, it's a journal.'

Sometimes I took notes, out of habit. I close the book.

'Why are you standing up? Why are you checking your phone? Where are you going?' Woman 8 sounds pissed off.

'I'm on call this evening.'

'Oh.'

'Sorry.'

I kept a log because I wanted to learn something, it's what I told my patients to do. It makes for shameful reading.

The previous year I had run the London Marathon. There was a festival feeling that April morning: thousands of children lined the roads near the start in Greenwich, street food, jazz bands, bunting on every house; it felt like the first day of summer. Never run on your feelings. But what to do with all this euphoria? I high-fived as many kids as possible in the first few miles, went through half-distance in sub-three-hour pace, was getting faster. I felt strong going round the Isle of Dogs, normally a cemetery for the underprepared. At Tower Bridge, among the Samaritans offering high-energy snacks, I spotted a woman in a sari handing out onion bhajis; I never could turn down Indian food. For the next hour I slowly choked, unable to work the bhajis – swollen by my saliva to twice their original size – past my oesophagus in either direction. Transient cerebral hypo-perfusion – from the lack of oxygen – meant that the cartoon-like crowd appeared to be jeering not cheering, the Dixie jazz band at London Bridge struck up Stockhausen, the children became hideous jelly babies bearing lurid, hallucinatory T-shirt messages – 'Cancer', 'Leukemia', 'Cystic Fibrosis' – the race turned into the hellish slow-motion denouement of a medieval allegory.

One thing I've learned: I don't learn from experience. I still hoped for something different this time.

The Literature tends to what Iron Man can *do* for you. A quasi-mystical package is promised, most obviously of health, a significant enhancement of physical but also cognitive faculties. There were smithereens of neurological evidence, as there is for everything these days. But there's a fairy-tale jump from minuscule anatomical findings to assertions that competitors will be more 'goal-focused', have 'sharper decision-making skills', 'improved time management', quicker learning rates, higher volumes (there's always a distinctly corporate undertow to the claims, of course). Then, more liminally, a communion with oneself that cannot be fully communicated to the uninitiated, the non-Iron (the Tin? The Cardboard?) but includes notions of self-actualising, repurposing, general transcendence ... In essence the Literature carries the not-so-tacit promise that once one has passed through its crucible, there is nothing that cannot be attained. *Iron Man.* Sounds good, doesn't it ... ? Would you believe me if I told you that these goals were not my focus?

What the Literature doesn't tell you is that it turns you into a moron: cognitively enhanced possibly, but still a cast-iron moron. Over the course of the nine months on my Swim, Cycle and Run Programmes, I stopped reading, I stopped thinking, I stopped dreaming. I put weight on from licensed overeating, balded quicker than normal because of swim-cap usage, walked like a rusted Iron Man – remain (several months later) unable to flex two of my toes, rotate my left shoulder or turn my neck to the right. Patients were sacrificed. I would finish clinical assessments in a rush to cram in a lunchtime swim: 'No, it's alright, madam, that's enough detail, I've got the *gist* of your problems. What ... ? No, it looks like a swim cap, yes, but it's actually a scalp conductor, from the team at UC Santa Cruz ... Got to run.' I became titanically self-involved. No news, no culture; Britain had broken from Europe and reverted to a nation of puppy-eating druids and I hadn't noticed. If it was my weekend I would stupefy the kids with fizzy sweets, iPads, endless loops of un-vetted films (notably *The Mummy Returns* featuring scenes of ravenous brain-roaches who ruined their sleep for months) which I would half watch from the 'plank' position or a stationary bike; or palm them off on Grandma. A selfish, irresponsible, deceitful moron. I 'bonked' at dinner parties – which I learned was a technical IM term for running out of energy. As for non-technical bonking, I never wanted it unless it was capped at heart-rate zone 2, which meant keeping the monitor on and sticking to a monotonous trudger's pace, however exciting things got. A sexless moron; an Iron Mannequin. And like the last days of Hitler, every week was the subject of incessantly futile, ruminatory planning to ensure ever-receding goals might be met, distances attained, metrics satisfied. A selfish, empty, neurotic, neutered, Nazi moron.

The sky was psychiatrically blue. I had the sense of waking up in someone else's life and as in the dream I had imagined for Murray, things had stopped working, the sense of something terrible having happened not too far away. Like the dream too I was missing my fingers, the water was so cold I couldn't feel them. My children were playing in the mud by the lake's edge untended. This is how the world would end. Wasn't this what the valley looked like? How could a dream I had fabricated turn prophetic?

What was I doing here? I suppose it was some kind of purgative for the suffering that had typically accrued over four and a half decades, and recently seemed to be accelerating. Then, more distally, the vicissitudes of the age itself: generation-wide restlessness, the sense of being distracted, dulled, frightened, under- and overstimulated, terminally lonely. These things were the real endurance events. Did I hope that this freezing immersion might provide some defibrillator-like jolt? These people were not my people. Watching them, shapeless black figures, isolated, clapping like man-seals to keep themselves warm, pointing with index fingers to the sky – semaphoring a god? Or placing him within quotation marks? I thought this explanation missed its mark: it was the pointlessness that was compelling: absurd, tortured, fated in a Sisyphean way, to grind out a life of slow-speed torment, heads bowed, steaming, heavy, warmed only by their own urine, like cattle.

Or water buffalo. 7 a.m.: the siren sounded for the beginning of the swim.

The Literature encouraged going against oneself, using strategically anti-impulsive mantras – 'Never speed up: never lose pace'; 'Tortoise-mind, tortoise-mind, tortoise-mind ...' – to make the scale of the race more like life; long, unpredictable, unwinding in unimagined ways. After a hundred-metre sprint in which I briefly led the field, my swim was swum: acute cramp in both feet, my lower back disappeared, a gallon of tarn in my gut, sinking. I was no longer cold because I was incandescent with anger; that what had eaten my brain for the last year was going so badly. I groaned deeply as though calling to a whale (in a landlocked Welsh lake) who might mistake me in my wetsuit for a sea lion, might then swallow me whole, might carry me round the entirety of the swim leg, regurgitating me over the finish line in first place.

'Picture it: one of those swish lifts in Westfield, jam-packed; me and half of Dubai airport in it.'

'So *racist*,' I say.

'Terrible, I know. Can you forgive me? The doors close, and then open again. "Room for a small one?" the guy says. In gets Samuel fucking-hell Jackson with his shopping bags.'

'You are kidding?' I say. 'This is a joke?'

Jokes can be important. He (Man 2) has this strange-looking R. D. Laing-style neckerchief, which looks like it's tied too tight.

'It is Samuel L. Jackson with his silly backwards cap. Everyone knows it's him, you can feel it, but nobody says a word. I'm right at the back, he's right at the front: it's a big lift. Samuel presses the close-door button. The lift door closes.'

He has a strangely animated face: it moves more than it needs to.

'Right at the moment the doors are shut I cut a giant fart, out of nowhere, a real King Kong-er. As the lift moves there's a ripple of disgust from those around me at the back, women in burkhas covering their mouths making hacking sounds.'

He is laughing at his own story.

'Everyone is holding their breath. Finally it hits Samuel, "Oh man ... fuck-a-dee-doo-dah!" – like he's been winded ... amidst all the pairs of eyes he searches out mine ...'

This is an anecdote: he is telling me an *anecdote*. I feel his performance's ambiguous movement; he is leaning towards me, compelling me to come and meet him, demanding approval, anticipating reward, and at the same time retreating behind the well-worn phrases, auto-piloting, elsewhere, like a speaking doll, its cord pulled many times before.

I want to get up and walk out of the coffee shop and be free.

'"Jesus," says Samuel, "what the fuck did you all eat?"' He does his voice with uncanny accuracy. 'He's pushing the 'doors open' button repeatedly. But the door won't open: "Oh maaaan!"'

'Hilarious,' I tell him, an ambulance *giving nothing back*. Really, I am already on my way home.

Anger then.

I knew that when my life became difficult, my response was to attack it more brutally, hoping to exhaust myself into faux-acceptance. But general, universal anger as well: we, these Iron Men and I – 'us Iron Men' as we say up north – divert midlife rage to exercise. Not a hobby, hobbies could no longer exist in this climate. This was an animating, monomaniacal seizing, a 'passion' as the Literature has it, with its unintended connotation of deliberate suffering. Drowning in this icy water had nothing to do with accomplishment or fitness or expansiveness or catharsis, had nothing obviously pro-life about it at all. Instead it was a collective expression of slowly unfolding hate, self-mutilation, disfigurement.

Why?

Because we, the unprecedentedly blessed, rich, healthy, mindful, happy, were also world-historically dissociated, unconscious and inert. If we were Sisyphus, then our rock was the self we lugged around. And though we might be dying inside, we weren't quite dead yet, because it was still a form of life to hate something (after Bollas 1993). I thought of Michael, whose *élan vital* was secretly built on a kind of disgust for all that kept him young. So too this Sisyphean labour in a blasted corner of North Wales was how we, the undead, might experience our aliveness: Flash Gordon and the Iron Zombies.

Well, maybe.

Somehow I had distracted myself through the entire swim. I tried to stand in the shallows but I had acquired hemiplegia. On the lakeside my children called on me to 'Hurry up!' and 'Stop showing off!' as I fell like Robinson Crusoe, face forward into the water for a third time. I crawled over sharp silted slate, made shore, posed for a photograph, had a slug of tea, took some of the laid-out High-Energy NutriPellets (aka jelly beans) that Bronwen and Cordelia were eating and then ...

I sat down and dried myself and had a second cup of tea, and a sandwich and a bag of crisps and a Twix and a meat pie, pork scratchings, a Scotch egg followed by coffee with six Tunnock's teacakes ...

'Get up, fatso.'

There was an edge to them these days, as though this was an abandonment too far. I had hoped they would take my absences in their stride, they had appeared unconcerned. But often self-expression is a waiting game with children.

'Excuse me, sir, would you happen to know the number for Childline?' they asked of a group of spectators, summoning their best Oliver impersonations. Team Benjamin would never be the same again.

I left the town of Llanberis on the bike and quickly found silence: a mind uninterrupted by others, tranquilised by a country morning in early autumn in Wales, set like a daguerreotype by a gossamer of frost. Just the sound of my breath, the feeling of my zone 2 heart beating in its cage and the warming endogenous opioid-tea that exercise *can* produce; the briefest moment when the homuncular-drug-DJ brews perfection.

This was what all of that effort was for; an emptiness on the far side of meaning, answering questions of who we are and what we are doing here with sweet nothing.

Emptiness won't stay empty. Anger again. For too long medicine has neglected the mind and worshipped the brain – the body's apex – as though the two were separated at birth. Nowhere is this prejudice more evident than in the so-called *Paralympic* Games. Which is why Lewis and I had extended the format in another one of our lunch-hour ventures. We looked at the possibility of legally advocating for people with mental health problems to be allowed to participate – part of a larger 'acknowledgement' drive. As well as competing against their privileged, fawned-over counterparts with physical disabilities, we would have events specially for them, targeting particular diagnoses – duathlon for bipolar, fencing for certified psychopaths – but also unveiling newly created events: marathon lying down for the depressed, ADHD decathlon (all ten events would run simultaneously), Stygian cleaning for obsessive-compulsive disorder. There would have to be therapeutic weighting to ensure dosages were more or less equal for different competitors factoring in the severity of the diagnosis, size of the patients, co-morbidities and other variables. This would hopefully tie in with major sponsorship from Big Pharma ...

'What work do you do?' I ask.

'I could tell you but you wouldn't understand,' she replies. It was only a few months ago, six at most.

'That's sexist.'

'It's racist: we don't like Yorkshire where I'm from.'

'Try me.'

'I am.'

She *is* different. Her face is quite unlike any other; open then blank then furtive then resistant then open again, within the same breath, so that I can only speed-read, missing out whole pages to keep up. Her hair is shockingly white, almost albino-white, white as flaming magnesium. Her skin is fragile, scarred by acne here and there, settling after years of turmoil. She speaks on the offbeat with a trace of a north-western accent – somewhere between Manchester and the Lakes I guess.

'And?' I ask. 'How am I doing?'

I can feel the weight of *her* scrutiny.

'So far – a solid four.'

Though all thought of hate and brutality are excised, there is still so much longing ... A hook in the gut reeled in since birth by God knows who or what. Crying, cycling, crying, fleeing something, feeling my way towards something else ...

Because wasn't The Brutal, weren't all these hopeless endurance sports really unconscious *quests*, for a Guinevere, a blessed Beatrice, or their real but somehow equally fantastical, equally unconscious counterparts – as yet unmet, likely unmeetable? An unrequested gallantry, an unsolicited sacrifice, a proof of meaning, a case for love, *the* cure. Only rather than pioneering life-changing neurodegenerative retardants to impress her, whoever she or he may be, or training wild white elephants to build beautiful giant obsidian obelisks in her honour, I was going round in poorly signed circles, guided by retirees in fluorescent vests and megaphones, with the lost and the lame.

And she, my Soulmate, wasn't even there.

'Four? Out of five?'

'Out of ten,' she says.

'What about your family?'

She doesn't answer.

'Are your parents alive?'

She won't answer. Now *I* feel nervous.

'How about I ask you some questions, Dr Strangelove?' says Woman 204, waspishly. 'Like: What is *your* problem?'

'Er …'

'What's *your* diagnosis?'

'…'

'Do you think it's attractive to be so *interested*?'

'Not—'

She won't let me answer. 'What about giving it with the intense look? All that nodding? I bet you're going to tell me that there's a really fucking strong connection between us, that you can *feel* it.'

'Ally, my friend – good to see you as always.' It was the Greek, from Cyprus, in the nick of time. We hugged like brothers.

'And who is this beautiful lady?' laying out before us half a dozen plates of sumptuous-looking mixed meze.

'Just another one of his fucked-up patients,' she says. 'You must be the wingman?'

'*Enchanté*,' says the Greek.

'And what's this "Ally" business? I thought you were Brad, Brad76 ... ? How could you lie to me? I thought we were soulmates.'

I liked her, Sonia85 from Clitheroe, near Preston.

Soulmates means something different these days, now that it's a dating website. Climbing the pass a second time, though the circling vultures above are evolving more quickly than I am cycling, I am no longer concerned by the race. I cringe thinking of that time, which was longer than I'm prepared to admit, when my only real relationship was with the shoddily invented self I projected in 'her' (and occasionally 'his') direction, a dimly imagined liberal, upper-middle-class virtual realm which I filled with various half-baked memes of perfect candidacy. And 99% of the time 'she' only existed as a sequence of equally poorly imagined symbols and algorithms.

First there was the name. 'Brad76' – anonymised like a case history, but for sale not protection, a homely mono-syllable, a Labrador of dependability, American-sounding, summer nights: couldn't be further from me dispositionally and six years younger than I actually was.

Then the strapline. I recalled earlier attempts: 'I see you are still having no luck catching me' rejected quickly for being unattractively immodest, and a Ripper allusion. Then 'How low can you go?' But there is almost zero tolerance of more or less open declarations of sexual proclivities on these sites. So finally 'You make loving fun!': more Americana, open roads, platinum sales, what could be less pessimistic, further from the real ... ?

In mitigation, the deathless, empty positivity of other people's profiles meant there were certain generic rules of tone and form that had to be obeyed – the self always being formed according to the demands of the culture – unless you *wanted* to sound insane.

Next the pitch:

Thank you, Iggy. Thank you, Elena Ferrante [*never read*]. Thank you, Arsène. Thank you, Thomas Magnum for your moustache [?] ... In gratitude for Sufjan Stevens's map of America, for Dwayne Johnson's smile, for Howard Hodgkin's sad, bright, bipolar palette[*I read it somewhere*], etc., etc. ... Without you, without you all, life would be so much thinner.

How could I ... ? How dare I ... ? The carefully selected stepping stones (high and low brow) of cutesy bohemian comfort, and the whole thing framed, casually, as an acceptance speech. Accepting what? A gong for a life well lived, with zero taste mistakes ... A code whose content was otherwise teetering on meaninglessness.

Likes:

Stella Artois [*sober for over a decade, but that doesn't play well here*], Radiohead, pomegranate salad [*puke*], wild swimming [*it's just swimming*], sunrise at Machu Picchu [*never been there but lots of Islington has*].

Dislikes:

Stella Artois the next morning [*the panic attacks, the tremors, an occasional seizure*], gastro pubs [*see the way I just turned on Islington*], parentheses [*Look, no hands, Mummy!*], earnestness but also disingenuousness; understanding there's a huge amount of leg room between the two [*pure bollocks*].

This list-ification of life was everywhere these days. We turn lists into stories, as though we can generate from splinters someone more complete than reality will allow. Meanwhile, in our loneliness, we dismantle our own stories, our selves that is, back into reproachful fragments.

I have good friends who mean the world to me, family is important, I like box sets, smiling dogs spinning water off their coats, getting your hand stuck in a jar of peanut butter [*or other abstract codifications of sex*] … etc.

The rubble of inclusivity, passwords for relative wealth, for mental health, the upper echelons of the normal distribution. As though the culturally acceptable cure for our pathological individualism – often cultivated in therapy's various programmes of self-enrichment – was meaningless pabulum.

And though the profile is a relentless stream of convictions, tastes, preferences, what strikes me as I now pedal back up the Llanberis Pass for what feels like the 10,000th time, is how little I've ever known about them, these preferences, any preferences; that they've never really felt authentically mine at all, just caught like flu, or dancing this way only because it's an amalgam of the people dancing either side of me. As change blind as the worst doctor, as suggestible as the worst patient.

Of course Brad76 could have written something more like the truth: 'Homeless, unreliable, ex-problem drinker, balding, middle-aged part-time father of two, permanently on the brink of catastrophe, as flammable as magnesium …' But where would that have got him in this brutal, Darwinian fight for a mate?

'Things you are looking for in a partner': 'lightness of touch', 'talks on the offbeat', 'lickerish-mouthed' – I might as well have put 'centaurs' … Another list of near-meaningless abstractions, half-attributes, held together by some childish fantasy that these variables might naturally co-vary, as though you can choose someone with actuarial-like specifications, rather than being guided and misguided by the crazed, butoh-like dance of feelings.

I was in free fall now, careering down a high-walled chicane at close to 90 kph for the last time, a severe traumatic brain injury waiting to happen. And the worse thing of all: half of the time I didn't know if Brad76 was being cynical or sincere. And if I didn't, how could he?

Back at Transition, Team Benjamin was in disarray. The girls – bored, crazed by sugar – were entertaining a small crowd of equally bored locals with re-enactments of scenes from something called *The Daddy Returns* – with jelly beans figuring as the neuro-ravenous bugs – involving multiple changes into different permutations of triathlon fancy dress to approximate ancient Egyptians.

They had eaten my remaining nutrition. That was OK because I wouldn't bother with the run. Sitting down on the cool grass as the sun lowered I looked forward to never getting up again.

Brad76's profile was *brutal*, an assault course of codes and doublespeak which has no way through it, which at least in part protects him from other people rather than attracts them. He was his own quest; its subject *and* its object. And it wasn't just Brad. See Fizz★, The Tomster, Gee72, Ems75 and their straplines: 'Take me to the river', 'Partner in crime', '24-hour party person', 'To die by your side'. Everyone was bang at it. Who were we kidding? Self-deceiving in order to deceive others better. Hundreds of thousands of us crazed over decades in myriad different ways by need, longing, loneliness, heart-break: the real *DSM*. And my final refuge seems to be in a form of addicting bloke-ish self-laceration, a tone which was nearing epidemic levels judging by the millions of words given to ironically depicted midlife crises (which, ironically, was the real though unseen crisis) in every paper, blog, magazine. This static, reviled notion of oneself was really just the obverse of a narcissist's self-idealisation, as though knowing one's deficits and passionately excoriating them might defend us against some-thing still more raw.

There was no soul in soulmates, without its constant dark night. I was on the site on and off for much too long before

I met Sonia, not consistently, but a month here, a fortnight there, serial offending for as long as I could stand it. I can say without hyperbole that the process of creating and perpetuating Brad76 drove me mad.

We were on the pavement outside the Greek's, about to go our separate ways.

'So, will you go out with me?'

'Are you seven years old?'

'Thereabouts, in some ways,' I say.

'Don't lunge.'

'What do you mean?'

'It's in your eyes.'

'What about a second date? Secret Cinema? The butterfly house at Syon Park?'

'Textbook Soulmates ... You're a handful.'

'How can you tell?'

'I'm packing mace.'

'Look, I'm serious.'

'No you're not, you're severe; chronic and severe. *I'm* serious.'

She is not blinking. Sonia, I think, like Raskolnikov's Sonia.

'This will go one of two ways; it depends on you,' she tells me.

The first stars had not pricked the skies above North Wales but blackness beckoned. It felt like a different day than the swim, it nearly was. No need to slow down any more, the run leg a crawl; blinded, fumbling Gloucester led by the madman to Dover beach.

Really we were all alone. The IM Literature talks about positive cognitions, that what you think about has a significant effect on neurogenesis during exercise, but at this moment I'd rather be accurate and despairing and have my brain atrophy than deluded and healthy. The berserk shoring up of the egotist's defences – in attainments, attributes, in tastes – had made self-reliant iron men of us all, meaning sharing, humbling, fragility, intimacy were impossible. We were alone. Or at least I was.

Quests lose sight of their objects, causes are lost, even self-excoriation softens as the passion dies: death emerges as the real and only end; love its mirage, its White Whale, its pantomime horse …

I'd been going for over ten hours and I felt like I was no nearer the finish line. I remember little of the run except that by the time I completed the first eighteen miles it was pitch black. I had to wear a head torch to get up the mountain. For the next hour or more there was the awful sense that I was being pursued, behind me a bobbing string of lights which stretched down the mountain drawing closer in silence, like a verdict, like abandoned dates, like poorly treated patients wanting to file suits. I no longer cared whether I was caught.

Below, the silhouetted corpse of Snowdon laid out under surgical moonlight, occasional clouds irradiated like brilliant hyper-intensities on an otherwise pristine MRI.

'Well, what do you think of that?'

My father's soft Welsh lilt nearly forty years ago. And I think about the closing moments of another endurance event, another ultra-marathon, less than six months before: the thirty-six hours by his deathbed, spectating a grotesquely protracted panic attack, wild utterings bound for silence, the face I know disappearing, greying then yellowing; and the long wait for the last breath, each one the same as the last except less, with the diminishing hope that we might voice something of the many things we had not begun saying. He didn't recognise me again.

The body changes in a moment: an empty encasement sinking back into the bed; a home that's suddenly stopped working.

What is the opposite of iron? Flesh? Breath? ... Keep breathing.

At the top of Snowdon I am checked by a doctor. I fake being alive for a moment while he examines me. There is a Literature on directed forgetting, the deliberate attempt to limit the future expression of specific memory contents. A form of repression, lying even – it's associated with similar cortical areas. But there are some lies that won't stay told; some directing that will not be forgotten. I am a patient and I tried to forget it.

My dad died and I nearly forgot that too. That was why I was here: on my father's favourite mountain, the whole escapade a pilgrimage, a funeral. I turn my pockets out and empty the rest of his ashes on the bare hillside, like I'd stolen them and wanted shot before being searched. I hear the steady breathing of my pursuers; wanting to tell me something else, one last thing, before I disappear for good ... Without knowing it, my legs are moving, have turned to liquid, to air, impossibly light, feeling no pain, no cramp, no rock, no shock, no earth – off-ground, never to touch the earth again; following them wherever they want to go, down the dark mountain to the tempo of 'Hard to Explain' (track eight – the Strokes); faster and faster, not a thought in my head, my eyes might as well be shut, just a hair's breadth of faith that stops flying turning into falling.

I cross the line with an absurd sprint finish in a little over fifteen hours, into the arms of Sonia, who's driven 200 miles as a surprise. I *am* surprised: she told me she never wanted to see me again when we spoke yesterday.

'My hero,' she says with chalk dust.

'It was all for you,' I say with more chalk dust, looking at the long-abandoned children asleep under a pile of discarded wetsuits, like freshly clubbed seal cubs. Sooner or later they would die. They already had on several occasions; falling into a swimming pool aged two, hundreds of road traffic accidents, a half-dozen toaster-electrocutions, knife fights, the 'toilet cleaner cocktail', the time little Bron's body stopped working … Why would I have those thoughts now? Who or what would visit such a thing on me?

It was over, The Brutal. It wasn't over, not by a long shot.

Dr Samuels

'So?'

'So?'

'Where would you like to start, Olly?' I ask.

'Wherever you like, Dr Benjamin.'

'Ally, please. Would you like a drink?'

'No.'

'You choose.'

'It's your session ... No, I don't,' he says.

'...'

'...'

'... You saw the hospital made the front page of the *Standard* again?'

He nods, not wanting to engage in my small talk.

'Deloitte have taken over the asylum.'

'...'

'...'

'...'

'I'm not very well.'

Occupational health referred me to Dr Oliver Samuels, a psychiatrist colleague. I had no choice. Actually I had had choices but I had made unprofessional ones: my clinics were getting shorter and shorter, I was dictating reports while the patient was still in the room, could be heard singing anthemically as soon as they left, their life-changing diagnosis still not dry; I pushed flexitime to breaking point completing my statutory 7.5 hours by midday; was in constant breach of dress code (I had developed the taste for wearing surgical scrubs; the feel of them, the leg room, the freedom and power they afforded). I had hidden things for a long time, dismissed the ache, ignored the limp, done without the crutches, explained away the wheelchair, made fun of the twin prosthetic legs like they had nothing to do with me – normalising and re-normalising each stage of my decline so that I still believed I might give those youngsters on telly a run for their money. Or I thought I had hidden them. Really I was the only person in the whole of Neurosciences who wasn't watching this slow-motion high-speed crash into an imaginary wall, wasn't watching because I was staggering round asking spectators what all the fuss was about, telling them how well I was feeling, naked from the waist down, half my cranium missing – metaphorically, you understand. Outside of work I had fallen off the property ladder a rung at a time – from two-bed, to one-bed, to studio, to shared house, and now the possibility of an unshared bed in a shared bedroom. Para-homelessness meant most of my time with the kids was spent playing 'I Spy' in London traffic in a disability vehicle I had bought at cut-price from a compromised Parkinson's patient (admittedly not all of these were strictly *choices*; the housing for example was in part a consequence of being on a public-sector salary and having an attachment problem which outlawed the possibility of shared living costs). But the apical bad choice, my magnum opus, was to get engaged to someone who was dead most of the time until she became terrifyingly, radioactively alive.

'What's wrong?' he asks.

I shrug my shoulders. I'm terrified of any exposure and, at the same time, that if I start telling him things I'll never be able to stop. What he's really asking is if I'm ready to dust down the old role, the 'patient'. I'd already been given some props by my GP: Citalopram and Risperidone to get precise, for low mood and disordered thoughts she said, but really they were memory pills to help me remember the part again. In this early scene I was sharing a room with a psychiatrist – unprofessionally (meaning *patient*-ly) – for the first time since my late twenties. My character is suddenly besieged by the realisation that that other psychiatrist was right back then after all. Which meant that the intervening decades, which occasionally felt like normalcy and freedom, were really just ecstatic high-speed flight, only to find myself and my pills waiting for me at the other end: pure Benjamin's syndrome ...

'*I'm not very well* ...' That's all it takes. The room instantly becomes his and I feel twice as sick as before I admitted it. I see him seeing me with his quick black eyes: my face a Baconian smudge, then an empty chair. I want to weep. I want to rest my head in his lap, for him to stroke my hair, to give the whole thing up ... almost. Still I resist, still I want a less charged scene in which the two of us have a professional-sounding conversation about a third person, a patient, who happens to bear my name.

He asks about my 'visual life'. He's good, effortlessly feeling out a symptomatic weakness of mine like a mosquito finding blood. As a child I was unable to draw the most basic objects, a kind of visual dyslexia. When I tried to 'see' something, other things would crawl into frame, interrupting my attention, so that I would want to white-out everything around the object. Last week, forty years later, I had a micro panic attack when I took the girls to the National Portrait Gallery. It was just a painted face, powdered, haughty-looking, lightly figurative, otherwise a nondescript eighteenth-century man. In the background was a table with medical calipers, a hammer and a saw.

He was a surgeon and a nobleman, maybe the queen's surgeon, quite possibly an ancestor of one of my colleagues. No wonder these doctors knew who they were, they had never been anything different, not for hundreds of years. I looked at it, waiting for 'him' to say something, but I couldn't settle on the painting at all; CCTV cameras high in the corner, the gallery wallpaper, the paintings and visitors nearby, all crowded my field of view. It was noisy for a gallery but all I could hear was the sound of blood drumming in my ears.

Dr Samuels had me draw in sessions and at home. When it was my turn to have the girls the three of us would spend afternoons in cafés, with crayons, felt tips, watercolours.

'What's *that*, Daddy?'

'Um, a sperm whale.'

'OMG ... it's rubbish.'

'I'm sorry.'

'Are you crying, Daddy?'

Thank God they had avoided my cack-handedness, their drawings free, spontaneous, alive, where I would labour for hours, in surprising amounts of pain, finding new ways to make each creature sink. If we don't learn to love our deficits we will only ever identify with what we *can* do. That's what I told my patients ... Some hope.

He asks me to visualise my 'Shadow', that least desirable part of the psyche which won't be digested. But I see only shadow: a gristly opaque mass sitting in my skull, a brain tumour the size of a brain, of a large grapefruit, its shape filling the cavity perfectly, making it hard to think as I thought about it. Why do doctors insist on fruit or vegetables as indexes of tumour size? Pea, grape, Brussel sprout, apricot, kiwi fruit, orange, lychee, guava, persimmon, cantaloupe, breadfruit – they change over the years, becoming more exotic with the metastasis of Whole Foods. There's something so doctorly about the cruel counterpointing of malignancy and health. Worse still, there's the knowledge that both things are alive, likely to grow bigger.

It is, he tells me, my job to talk to the Shadow, to care for it.

'Darling, I'm here for you,' limply following his script.

'Once more with feeling,' he encourages.

'. . . Get-to-fuck, soppy gobshite.' He was finding his voice. He had a Glaswegian accent, as tumours would.

He tells me to keep loving it, whatever it says.

'You have to believe in change,' he is telling me . . .

Change: fine word. Neuronal change is instant, neurotransmitter levels fluctuate in microseconds, whole synapses can come and go over the course of a long lunch. But at the level of networks – of behaviour, minds, people – it depends on long-term action, is slow, incremental, often circuitous. I so desperately want to believe that change is possible, that we are not fated, like stories, that anything can happen at any time. But developmental pathways *are* predictable, especially if there was trauma at a very early stage, before language and visualisation had a chance to take hold – then we might expect a bias towards repeating old traumatic responses, again and again and again. Change, but really no change. There are only two outcomes. My work is based on an assumption that change is possible, too often my life on the assumption that it isn't.

He suggests I write to myself as if it were twelve months from now, write from the other side of this current storm.

'The letter might be a fund of hope, of hard-earned wisdom, to be drawn on in moments of fragility, a promise of changes already made, of imagination, of faith. Write, literally from your frontal lobes to your limbic system, Alasdair.'

'It's not literally; brain areas don't write, or talk or think or feel; people do,' I don't say. *Literallys* – they were everywhere, used thoughtlessly as intensifiers, but also as anti-anxiety drugs against the inflationary panic of language itself. But if we pop the pills now, what are we going to do when the world really heats up? Literally.

> ... you who will get lost, fall victim to the "inner police", to shame, who will believe too often in the apparent facts of what you are feeling, or worse – that you are unable to accept any help that living itself might offer – is the selfsame you that right at this moment is perched on such good feeling, such potency, peace, purpose, Pimm's, piss-flaps, Portugal, pappish, pilates, pudenda ...

I can't do it ... A dozen goes and my future self sounds so manically, mechanically happy, more pathologically fake than the self it is trying to rescue, drowning him in hope. Rather write a response in reverse, from the fractured patient of the future to the present's smug-sounding doctor (me that is), thanking him for nothing, asking to be left alone, threatening a malpractice action.

> Dear Dr Samuels,
>   I am your patient. More than anything I want to give you what you want. Diagnose me with a disorder of your choosing; I will do my utmost to behave accordingly. Even if you demur I will read your mind somehow, give you what I think you want by your smiles, your encouragement, your disapprovals, even when they are not meant, even when they are not there.
>   Your patient until the end of time ...

When things fall apart the reflex is to scrabble for the type of security that will only extend the pain schedule. So I find a ready-made substitute for the patient role in my fiancée; she would provide me with direct competition in the insanity stakes: joint favourites in the Paralympic codependent three-legged race. Though undeclared in our profiles, this is why we had 'chosen' one another as Soulmates: at the time I rationalised that it was her offbeat spikiness, her northern-ness, her lickerish mouth, blah, blah, blah ... really my Shadow instantly recognised hers scrawling me love notes in lipstick on a dark mirror. It's not long before our more superficial self-presentations flounder and we both clearly see we have a bona fide patient on our hands; unmanageable, unaccountable, in free fall. There are bad fights; thrilling reunions; worse fights.

I resort to a desperate gambit in my next session with Olly (I alternate between 'Oliver' and 'Olly' depending on what I want from him):

'I know it's not standard practice, Oliver, but Sonia would feel comfortable around you, I know it. The situation's acute, it demands something creative ... She's not straightforward [*like I am*], and you have such a gentle, amiable, imaginative way [*oooff, what charm*] ... I can easily find another clinician for myself [*schmooze my way out of the professional help I need*].'

After clearly stated reservations, Dr Samuels agrees to stop seeing me and start seeing Sonia immediately, on the condition I find someone else.

Within two sessions Sonia is refusing to tell me what she and *her* psychiatrist speak about: 'That's between us.' Then she tells me that of course I've come up.

'That was between *us*,' I tell her. 'What about my confidentiality?'

'That's why I can't talk to you about it.'

'We're engaged. We're supposed to share everything.'

Instantly I am jealous of both. I wonder if Olly prefers Sonia to me. I wonder if Sonia prefers Dr Samuels to me.

But it's much more than jealousy, I rationalise. It is clinically important that Oliver and I share information, have an *ex cathedra* discussion about the health and prognosis of his new patient, my future wife: for her sake, for mine, for ours.

So when I bump into him on the Stroke Ward, and again in Electrophysiology, and for a third time on 'Salad Island' next to Bariatrics, I see it as professionally necessary to mention something about Sonia's symptoms.

What I'm worried about – and I've seen this in numerous patients of a particular character structure (the distancing jargon of authority coming so effortlessly) – is a propensity for mindless exaggeration, sometimes fabrications that are chillingly wild, and likely, I believe, to have their basis in a potentially malignant narcissistic aggression ... For instance, I knew instantly that she didn't approve of a pair of flamboyantly striped socks Cordelia bought for me as a present. When they went missing Sonia flat-out denied having seen them, looked nonplussed, said it was a real shame given how well they suited me. Three days later I find a shred of the label stuck under the blade of the NutriBullet.

'You know I can't discuss this with you,' Dr Samuels tells me.

Yes, I do, but all I mean to say is that there may well be a danger of the real being mixed with the unreal, inside with outside, thought with speech. Risk might be dangerously minimised by her, and, not knowing her, by you.

'I insist you stop right now, Alasdair, this is not what we agreed.'

I just want to give you a thorough handover, Oliver, so that time isn't wasted pruning imaginary gardens.

'Enough! ... Have you found a new therapist?' he asks.

I have finished speaking. I have nothing further to add. I am interviewing for new therapists as we speak ... But if you needed one symptom, Oliver, just a tiny one, to illustrate my point perfectly, then I would direct you to her laugh. Just listen to it ... Superficially it sounds authentic, doesn't it? But you'll soon come to notice how it's never quite free, that there's always something ulterior mixed in, a little hook punctuating some raw, malicious urge ...

'Shut up!'

'What … ?'

'SHUT UP.'

'What did you just say to me?'

'I told you to shut up!' she says. 'So shut up!' Sonia starts to laugh. I am not laughing.

'Your voice is dangerous … Shhhhh.' Sonia putting her finger to her lips, hiccoughing with laughter, the sound of it climbing the 200 metres to where I wait, halfway up the pass, three weeks later on our ill-fated trekking holiday. I'm the butt of some joke she is making with the guide – a local in his early fifties whom we are to call 'Kid' for some reason (a young moutain goat? But he's old, bloated-looking, hamsterish more than caprine) whose comprehension appears limited though he talks fluently in this irresistible mix of Japanese-English, his two main tourist constituencies. She's been like this for the last hour: light-headed, giggling incontinently, then stopping suddenly. She appears totally unconcerned by how far we still have to go if we are to make the tea house before nightfall. I know different.

'If you're not quiet, you'll cause a landslide with that mouth of yours,' she shouts.

'I didn't say anything, it was barely a whisper.'

'Exactly.' I never could whisper.

And she's off again, triggering Kid like an ecclesial call and response, their guffawing arpeggios climbing high above the steepling valley wall. I have always suffered at the sound of other people laughing or eating. It means I am not doing either.

I walk ahead to anticipate danger – is what I tell her – to mitigate it where possible with encouragement ('The next bit gets much easier') or distraction ('Keep your eye out for big cats'), knowing that either could just as easily provoke. Sometimes I leave a small, hastily constructed heart of pebbles, or a cairn of cashews and figs – her favourite snack out here – for her to discover. But, really, I walk ahead because of the less selfless need not to have to hear that laugh, or look at her, or at Kid who is obviously struggling with the climb (I didn't want a guide, but I thought it might pacify her, or at least divide her resentment), or at anybody else for that matter, offsetting for a moment the relentless escalation of anxiety with the deception that I am alone, a dog out front, a dog to her cat.

I could count my blessings. This is the first time in over a week – her first week in a developing country – that I've heard her laugh; subdued as she has been by the toilet-less, chilly, incessant sawtooth purgatory of densely forested, pathless valleys and gorges – barely a view to savour – our long lead-in to the more majestic snowcaps that are our goal. The walk had been recommended by the standard guide as a 'challenging' but 'anthropologically interesting' alternative route to other easier, more popular ones. But there is hardly anybody here, 'anthropologically bereft' is more like it; a string of small, semi-derelict hamlets on impossibly farmed hillsides, or pasture with a few famished-looking livestock, empty, stricken monasteries still decked with brightly coloured prayer flags, as though something had gone terribly wrong at a children's party. Or a wedding. On top of which were various 'Third World problems', or 'First World Third World problems': insects, water quality, food hygiene, the trustworthiness of the locals, alongside 'Third World Third World problems' like the possibilities of flooding, rabies, avalanche, earthquake. More than once I regretted that we hadn't just gone to Snowdon, but it was me who insisted we come here – 'It will be stunning,' I tell her. 'It will make or break us,' she tells me. Really it would *stun* us and then *break* us. More really: we were already broken.

Even though I am fit from all the IM training the going is particularly hard with 25 kg on my back – the weight of our combined rucksacks – at this height (3,000 metres approximately); I can imagine what it might be like for her. I know very little about the physiology and practicalities of high altitude. Ideally we shouldn't ascend more than 500 metres in a single day; 800 metres is the upper limit, but 300 metres is preferable. Frequent breaks, regular fluids, and still to expect bouts of dizziness, nausea, shortness of breath obviously, with the spectre of catastrophic pulmonary and cerebral oedema for the reckless or unlucky. I don't know what she knows about it, but I'm guessing it may be a blind spot in her otherwise postdoctoral-level knowledge of physical health (she was a medical lawyer by profession), of negligence in general, reflecting a need to subcontract some small aspects of life-management to me in order for her to feel connected.

Today we have to climb well over 1,000 metres, otherwise we won't make it. Pushing the pace, trying to drag her up the slope by force of will, I put my foot on a rock made greasy by the drizzle and fall heavily, cutting my ankle on a root. That laugh again, erupting with my fall, echoing high above the rocky face, sounding more shrill and poisonous than usual, as though stored for days, waiting for its moment. I can feel hot liquid flush the joint; I diagnose it as broken. I'd always thought of myself as having a high pain threshold, so I'm disappointed when I pull down the sock to see a small cut on my ankle, incommensurate with the pain. I pull it up again as she arrives.

'Have you hurt yourself?' she asks still sniggering. Kid helps me to my feet.

'OK?' he asks as I shake the ankle out.

'OK?' she asks. I can't look at her.

She laughs one more time: I know there will be hell to pay. The dependent person resents being enslaved, because they've asked for it. This was exactly the kind of thing I had tried to warn Dr Samuels about whilst really heedless of it myself, because I couldn't fully give up being the clinician, as though the danger was hers and not ours, as though she could manufacture the explosives entirely on her own in her high-fenced laboratory.

★

We leave at dawn the next morning. It's a beautiful day. I'm wearing shorts, a T-shirt, flip-flops. She's in an unlikely red summer dress. I've always reacted against the genteel fascism of trekkers (what's wrong with 'walking'?) and their 'kit'. She takes my hand, I feel how light she is. We walk up through a pine forest in silence to reach a ridge that hugs the side of the valley. At last a view: it is Snowdonia or the Lake District in early October at the eyeline, only above does it become ludicrously different, a white-capped War of Gods, as though still carrying the force of the terrific collisions, the unimaginably terrible fury that had formed them. And it was the echo of this, along with their inhuman scale, that seemingly subdued the possibility of conflict between us.

A little later when I stop with the guide to take a piss, Sonia is actually skipping, singing and skipping. By the time we finish she is a red dot 500 metres further up the valley. In several moments of exuberance over the last few days I had challenged Kid to a race, comically at first, then, sensing his lack of fitness, his rheumatic-looking hips, his pudgy frame, with a slight edge. But now, sniffing his moment, he pats me on the back of my pack: 'Race you, white boy,' taking off with improbable lightness and speed up a steep-looking crag; suddenly he *is* a kid, a youthful goat released into his natural habitat. I'm more than a little surprised that my dull legs, which recently carried me to a top-ten finish in the World's Toughest Iron Man, don't instantly respond, as though they're thousands of miles away from the brain that ushers them. When they do, they feel like they're moving through fluid not air. By the time I catch up with Kid and Sonia – they have had to stop and wait for me – I have a cracking headache, my vision is blurring, my footwork is drunk, my heart a thousand angry bees banging against a window. Or rather I have decided that these symptoms are there and they have obliged me.

'You OK?' she asks, checking a map. 'We should have a quick breather and get cracking. Need to push it if we're going to make it by dark.' All of a sudden she's the leader, checking on the troops. 'Does that sound good?' she asks rhetorically. I think she would not be here if it wasn't for me.

Then I tell her I have a mild headache.

'Take it steady then from here. We must be at four-two now?' turning to Kid: Sonia – the mountain guide. She removes her fake sunglasses to wipe her face. The black dye from the plastic has left a deep stain around her eyes.

'Chinese panda,' says Kid pointing at her face, laughing. She looks at herself in the mirrored glasses and that laugh starts up again.

Walking away I instantly regret admitting to the headache. Unbeknown to her I had spent almost every minute of the last ten days fretting on her behalf, worrying about the time she was having, a moment-by-moment calibration of her mood. Not easy given that her face was as enigmatic as a cat's. Or it was legible but unstable, like the local weather: appearing to smoulder, until brilliant sun broke through, or just as easily, a terrifying electrical storm. 'What sort of a man ... ?' was the prefix to many of her attacks. From her mouth the question took on a genuinely non-rhetorical, typologically curious meaning. That was the killer irony; because however reckless, or ungallant or disorganised I might be, I was super-sensitive to danger when her feelings were at stake. Distracting her with questions so she doesn't fully appreciate how cartoonishly precipitous the drop is when the bus tears over a barrier-less, narrow, pockmarked pass; consolidating our rucksacks so that I might carry everything, knowing that any additional effort might push her mood off a ledge; like waking her with breakfast every morning – but only when she felt she'd had enough rest – fried flatbread with honey and peanut butter, and English tea which had taken a day and a half to source; like spending an anxious hour searching for insects by torchlight, hoping to secure each night's bedroom before retiring, anxious because knowing that to miss just one would for her be catastrophically symptomatic of my carelessness ... And on it went for almost two weeks, the low-level panic attack of caring for someone.

Further and further ahead now, staggering but fiercely determined, brooding on how almost every potential blow was cushioned, every setback or fear; most of them not even noticed by her, not once have I veered from patience, diligence, concern, without a flicker of appreciation. (If I thought about it I was treating her like she was a child, my child, treating her better than my children in fact because I was over here, 5,000 miles away, rather than looking after Bronwen and Cordelia: the abdication of my real responsibility made me wince.) Anyway, the first time I mention something about myself, some tiny symptom of the fact that I am an organism outside of her, I am mocked, dismissed, laughed at ... Self-pity is addicting. I concentrate hard on the headache, the heart rate, the reduced lung capacity, the weakness in my legs. Noticing begets noticing: this is the law of attention.

I am not well. Sitting in a deserted tea house I tell them I can't go further today. The spell I have cast on myself is complete, almost. It's hard to admit, but I still have a quarter-eye turned outwards, to how I might look.

She puts a plate in front of me. 'You should eat.'

I shake my head, a prelingual child.

But if anything will soften me, it is food: dumplings, apple pie, sweet tea. A huge silent relief that is stillness passes through me like sun, a warm tongue on the cub's crown, the thought of being thought of. Breath becomes easy. Soon I am lying next to her in bed, both of us in hats and scarves, under five thick blankets, Sonia stroking my face, singing something I think, ourselves again: I have never been in love like this before.

I had just about fallen asleep. I still can't remember what momentary coolness, what micro-sleight sparked things up again. On her side: a spider's shadow, the distant rumble of an avalanche, or just the fact of me falling asleep first, turning away to face the wall; on mine: her refusal to have sex perhaps. Whatever the truth, the fragile truce is shattered. Within moments we are soaked in each other's petrol, wielding giant flamethrowers: 'What sort of a man … ?' She cracks me in the face with the flat of her hand. Shocked, furious, I push her out of bed. Grabbing her things she runs down the dormitory corridor and locks herself in another room.

After five minutes, righteous fury has turned to panic. For the first time in years I try to meditate, but focusing on my breath only accentuates its jagged shallowness. I am going to die. With the help of the wall I walk down the corridor to the room she has locked herself in, each footfall a deepening of role, step by step, misattribution by misattribution, a diagnosis in motion. I knock, whisper her name. There are other trekkers staying either side, happy-looking couples from Italy and Germany we'd watched from a distance at lunch.

'If you don't open this fucking door I'm going to kill myself.' I just can't whisper. Still she doesn't answer.

It's minus 20 degrees outside. The night sky, close as it's ever been, is lesioned with stars, a billion tiny, sparkling malignancies. The mountain has moved nearer overnight, bending the space around it. I think there is something glacial, unknowable and dead in all human relations. Holding my breath, wearing signature shorts, T-shirt, flip-flops: I close my eyes and walk. Do I really want to die sulking? Will I have a choice … ? What about the girls … ?

The doctor removes the oxygen mask, asks me to get off the bed, and to try once again to walk in a straight line. I can't do it. He thinks I'm trying, but what does trying look like? I've always been clumsy, I tell him. I can see he thinks I'm the brave, stoic type. He places a clip on my finger to re-measure my saturation levels and makes another mark on his chart.

'I'll give you another dose of steroids, OK?'

'Really?'

'You're still in the moderate-severe range.'

'I am?'

'It's often the fit ones who get into trouble. But it's usually young bucks, you should know better ...'

At some point in the fugue a sober thought must have broken through and made the decision to turn around. I found the hospital easily enough, not too far from the tea house: a recommissioned temple and a couple of tents, set up a few years before, the charitable idea of another British doctor who had seen enough of poorly looked-after porters, often unskilled labourers from the lowlands, unused to the altitude, and too often left by the roadside to die by heroic lawyers, accountants, doctors et al. pushing on for their bucket-list views.

'I feel quite a bit better, Doc,' I say.

'Tough guy, huh? Listen, you have to descend now the light is good enough. OK? It's not safe for you to go down alone,' he tells me. I tell him it's not safe to walk down with her. Me: 90 kg, barrel-chested, race-fit; I should be ashamed of myself. I was, I still am. As the doctor mulls over the options I notice for the first time the poster that says 'Charges'. Together the steroids, the Dimox and the oxygen and the overnight stay will cost $1,600. The low-budget trekking holiday suddenly has a five-star-honeymoon feel. I had asked her to marry me after less than a month of dating, I hadn't meant to but I couldn't resist.

I hardly notice her walk into the consultation room. She has been listening outside, heard me say I don't feel 'safe' in a gentle professional tone, deferential to the doctor yet quietly decisive about my symptoms. I know she has heard because she is walking towards the doctor telling him that she is here to give a 'collateral fucking account', here to tell the doctor in her thickest Preston accent that there is nothing-the-fuck wrong with him, that it's all in his fat fucking head, that he is made of shite, that he has psycho-fucking-logical problems, that he is a liar, that she doesn't give a flying fuck what the fucking 'obs' say, that he's always been a lady-beating liar, that he has a history of this sort of thing.

And I remember thinking, even in the midst of it, what a talent she has for swearing; not like the other girls I'd dated – half-hearted, fey, cutesy – but nosiy, spit-full, acidic, *meaning* it. Then I wonder why the doctor is not telling her to get out, or asking for my consent, or writing any of this down.

'Did he tell you he's a doctor? Get-to-fuck he's a doctor. He's a psycho-therapist, that's two fucking words in his case. You see my face?' – she points to the dye from the sunglasses – 'He broke my nose a few days ago, that's right, the fucker headbutted me.'

'The fucking-lying-bollocks I did,' my Shadow-tumour is saying, 'you cannae believe a word this witch is sayin',' in full Glaswegian. But nobody is listening. I think about how, despite all my efforts, I failed to predict this earthquake; whorish intuition had deserted me.

The doctor is still not writing any of this down.

'Last night he raped me; we had an argument and he pushed me out of bed because I wouldn't have sex with him to calm him down. So he raped me.'

Finally she turns to me:

'Tell him … Tell him the truth, you fucking madman.'

The truth? I'm the one who feels like laughing now, even as she's bawling at me, remembering how in the middle of that hideous, sleepless night just gone – a night I didn't think I would survive – I had, between performances, been half reading one of the medical leaflets for the umpteenth time, which explained how sometimes altitude sickness is associated with rapid impairment in learning and memory and, on occasion, profound confabulation (neurological 'bullshitting') because of the effects of atmospheric pressure on the frontal lobes: she is the one with real mountain sickness. Either that, or her character just took another dark turn on its axis.

We head back down the valley, three of us, Sonia a long way out in front, carrying her rucksack for the first time in two weeks, the guide in the middle, skipping forward to keep an eye on her, then hanging back to make sure I haven't collapsed; with me keeping up the rear, keeping up the diagnosis, keeping out her charges. But really I'm feeling better as we descend and the air thickens, just as the doctor had said. I wonder if Dr Samuels agreed to treat Sonia because he knew I couldn't leave a relationship that was evidently hostile, so he switched to working on the other half; or, I wonder, had I unconsciously done the same thing and transferred responsibility to the therapist for a decision I was too frightened to take; or, I wonder again, had Sonia projected that same wish into me, and now into Dr Samuels, to embolden herself vicariously? Or even, if were we all poorly drawn figures in Kid's dream, to be woken from by this day's end?

'By the way I've called the police,' shouts back the blood-red figure in the distance. 'They'll meet us when we get to town.'

I don't care about the police. She is my love, my Binding Problem, my Shadow, my tumour: I can't remove her without destroying myself ... Thinking about it, though, even if they don't charge me, I'll still have a record, my professional body will require me to notify them, I will almost certainly face a tribunal, it's possible I'll be struck off, that's if I'm not in a filthy, featureless cell for months, years even, waiting for a trial where, unable to afford legal representation, I will have to defend myself in a language I don't understand, in a country where there may well be a death penalty for rape ... I think of that other Sonia, Raskolnikov's girlfriend, the prostitute who tends to him daily in his penal hell, who saves him with her long-suffering love ...

This Sonia would have her own story that was nothing like mine, and that's the only story Dr Samuels would hear. Really I didn't know who she was, even in my imagination: all the alienness, the paranoia, the aggression that I attributed to her might just as easily have been mine. And who was 'I' other than a hopeless guess at what the 'she-I-couldn't-know' would want from me? As lost as each other, trying to hit moving targets from moving targets, neither of them really there.

We are back among the trees, the oxygen-rich, the Wi-Fi. By the time we reach the European-style café she has softened too, as she has so many times before in the few weeks I've known her. There is no line which is uncrossable, in either direction. I read about my team's latest loss, send a message to the girls about my thrilling encounter with the abominable snow-woman ('*just kidding*'), as Sonia uploads her photos for her family in Lancashire; 'One less bucket!!! Gonna need another holiday to recover from this one!!!'

I think of asking her about what went on back there and think better of it; likely she has already forgotten. Instead I stroke the back of her darling neck which arches at my touch and she lets out a single shriek, a note of glee, laugh-concentrate: she is a cat, I am a dog. She is a cat but she is no longer holding my trembling body in her jaws, or fixing me with almond yellow eyes, only purring and padding.

'I feel like you really understand me,' I say.

No reply. Dr Samuels still doesn't look like he gets it. Instead he looks frightened.

'You see, even at her most peaceable, that laugh, that involuntary giggle – well it seemed to bring on the next lot of damage. That's what I was trying to say before we went away ...'

I hadn't meant to tell him the whole story when I cornered him again in the General Neurology corridor a few days after I got back, just the bit about Sonia's laugh. Did he listen without interrupting because he was frightful I'd make a scene?

As I told him about it all – framing it again and again as critical information about his patient's symptoms – I thought about how he would hear the whole thing from the horse's mouth soon anyway, if he hadn't already, which made me say more and more and more, defending myself against every imagined charge. In that way I accidentally buttonholed him for a full ninety minutes in full view of a packed waiting room, sucked into the story of a breakdown that was as much mine as my fiancée's: textbook Benjamin's syndrome.

But whatever the rights and wrongs of this professional hijacking, my instincts about Sonia lying were right. A few weeks later she called and casually let me know that she'd had another boyfriend all along (I thought I could hear her stifle a giggle). She had allowed me to support her while she looked after him: he had one kidney and a stomach that couldn't break down solids. It would come as a surprise to Dr Samuels, if he hadn't known all along. But though vomitously painful, it wasn't a surprise to me: I well know that the patient will always finds ways to escape, slip the ID bracelet from his wrist, put trousers over his standard-issue pyjamas. Me, you, Sonia, we are all at it. Even the unwell boyfriend would have someone, no doubt – a blind mouse, an ailing plant. On it goes like an infinitely nested doll; however sick, however mad, there would always be someone worse in need of looking after.

You again

Alone once more: a lost, mate-less soul.

Except not quite alone, a thud behind me when I walk home, the sound of someone breathing at the foot of my bed, the familiar sense that something terrible has happened and is about to happen.

I spend time with Jonah. My work with him a parable of clinical futility but distracting. We meet in cafés, drink coffee, smoke cigarettes, we eat falafel, we go to watch Leyton Orient FC. Our second team. My third team. We take each other's tablets. He gets his ear pierced. We talk about our daughters, I think; him in Creole, or pidgin, or Arabic (absorbed from Al Jazeera?); me in English.

Illness separates us, there are two ways to reduce the gap; either the patient gets well, or the doctor goes downhill.

A week later on a summery late-September morning, Jonah flashes a carving knife at a new neighbour, a feckless-looking student, tells him he will chop out his gall bladder and feed it to his pet soldier ants if he doesn't leave the building. Jonah tells me about it after, in broken English and charades. I make a note. He laughs. I make another note. He laughs harder. He makes a note. I laugh. I am not concerned: he's a natural showman, he says these things for effect, we all do. There is risk here, whatever its severity, but I have minimised it. His curses are phatic: it's just like talking about tropical weather to him.

We ride the Circle Line. We eat halal burgers. We get our hair cut. We go to the dogs in Walthamstow. More than distracting, the ordinariness of it is an antidote to all that elaborate, evidence-based knowing, all that *expertise*. A two-way intervention. We play blackjack in the Turkish bath under York Hall. I get my ear pierced again. We hold hands as we walk down the high street. It's autumn in London, but it's the same temperature here as Port-au-Prince.

They section him. I'm given a caution for not alarming the relevant services earlier: a panic button that makes no sound when depressed. I visit him on the locked ward. He says the CCTV cameras they have here make him feel at home. I give him cigarettes and 'full-fat' Coke. He uses it to swallow a fistful of tablets.

'They told me the tablets will cure my brain injury.' The things these guys say!

He's upset. I'm upset. He can't understand what's happened to him. He doesn't know where he is or why he's wherever he is. I don't either.

He more than any patient before or after – exotic, chaotic, traumatised, estranged – is my imagined other: I want to lie down in bed next to him.

Things had got worse at home. I was staying in friends' spare rooms. I was running out of friends. I could only sleep in hour-long stretches. Most of the time my heart was stuck in zone 4, on the verge of attack. I convinced myself that I had fallen and hit my head and had lost the memory of the fall: this was the only possible explanation for the blurring of vision, the tinnitus, the persistent aluminium taste under my tongue.

It took me ten minutes to button a shirt. My signature changed without my authorisation. The basics had become tricky.

They couldn't get rid of me, not with the hospital under so much scrutiny; you can't be seen to discriminate against the unwell when your department specialises in mental well-being. They stopped my clinics, gave me pointless audits to complete, sent me on risk-management training, made me in charge of information governance which meant tidying the test cupboards. But at the same time as I was forbidden patient-contact my clinical instincts, usually half buried under a pile of preoccupations, were sharpening exquisitely. I could hear the squirrel's heartbeat. I could hear the lion's roar. I could feel the feel of different neuroanatomical regions as they acti-vated. I mean I could feel my brain in the same way I could my foot or tongue. I walked over London Bridge in rush hour, faces thronging around me, and diagnosed each one in an instant: *Psychosis … Depression … Lewy Bodies … Panic … Depression … Sociopathy … OCD … Cynophobia … Panic … Guam's …* Everybody has something, and now there's a name for it, even if it's fear of having something, of going insane, aka *dementophobia*.

They must have forgotten to rebook your follow-up appoint-ment with a different clinician. Here you are in my room again. I knew you'd find me; the timing couldn't have been better. Has it really been nine months? God you remind me of Helen. You've had your hair cropped so that it sits well behaved on your head like a beautiful, ornate silver helmet. I've thought of you many times since we last saw one another, struggling against the tides of your life. I thought about calling. I thought about taking your temperature with my tongue – I couldn't help it. The French have a phrase for it: Médecins Sans Frontières: Doctors Without Boundaries. If only we'd never met like this, if only you'd not stood up when I called out your name, if only you'd not crossed the threshold of the clinic, we might have found each other another way. I checked online: my regulatory body says that two years must elapse without professional contact before romantic relations are permitted. But we don't have two years. You don't anyway. From a distance your face looks calm enough, but up close I can see minuscule spasming in the palatal muscles, hear a new nasal timbre in your voice caused by lower motor lesions in the medulla oblongata, I imagine. This is the beginning of your future.

Things had got worse at home for you too. Your father had died, your husband had moved into the room downstairs, your daughter was in and out of hospital. You no longer drive because of one too many near-misses. You have labels on the outside of the kitchen cupboards to remind you what's inside. You can't make sense of your accounts, the numbers just slide down the page. You're more and more unsteady on your feet. But it's still you alright, still more alive than the rest of us put together.

'Don't tell me,' you say. 'Screwdriver ... Belgium ... turquoise ... grapefruit ... ?'

I don't want to test you again.

'Have I remembered it right?'

'There is something I have to tell you,' I say.

'You don't have to tell me.'

I do.

'What's wrong?'

No more hiding behind writing.

'You don't look yourself, Doctor.'

More than you could know. I have not been completely straight with you, with any of you. Some of the patients were me, are me – Junction Box, Bede, Craig, Benjamin – at different ages, in different stages; the other half were you – my patients – with me as doctor. Everything actually happened to me as I described it; the self-electrocution, the overdose, the flashing, the monk phase, the Ripper fantasy, the deranged diagnostic diagrams, etc., except I have made time lapse inconsistently, have changed things for the purpose of confidentiality, and also, sometimes, to make me or my circumstances look a little better or worse. The changes I am responsible for were made in good faith, apart from when things just changed without my say-so because that's what our brains do, whether we want them to or not; the deception is part of my condition, part of our condition.

'I'm sorry,' I say.

'These things happen ... I'm sure you had your reasons.'

I did: I hoped by allowing my imagination to spread itself slowly through these lives, through this life (like a disease, but really more like health) the truth might emerge more fully. Staging things like this and then writing it up was the only way I could think of to bring the very different strands of myself together, bring them under some kind of professional scrutiny, if only my own. I wrote in the hope I might create a doctor who could care for me. And in writing it I realised that they're not exceptional stories, just exemplary ones; if they didn't happen to me exactly, they happened to someone else, you can bet on it.

'Can I get you a glass of water?' you say.

Only writing it out hasn't quite had the stabilising effect I'd hoped for. Too often I failed to recognise the patients as my own, or let myself be recognised by them. That's what shame can do.

You are peering under the desk where the real Dr Benjamin has been cowering, as exposed as Tom-a-Bedlam.

Unaccommodated man.

'You're in the wrong career; you should be on the stage.' And you take me by the hands, and stand me back on my feet.

'You'll be fine in a moment.'

'Will I?'

'Of course you will. We're all just dancing, making it up on the spot.'

Unsteady yourself, you bring me close and the two of us are leaning on each other, trembling, like Fred and Ginger in their eighties. Meanwhile the TV crew have just finished setting up for a single long elaborately fluid tracking shot.

'Now, let's see what you got,' you say.

'Action.'

The camera takes in the sign outside the door which reads 'Quiet: Assessment in progress'. Radiohead's 'Nude' is playing on the soundtrack, a brass band arrangement, with Hank Williams on vocals (for it was your fantasy as well as mine). We dance under strip lights, tiptoeing around cheap furniture in the small office, humming our own accompaniment. I'm lifting you and your disobedient legs off the ground and into the air, round and round until at last we are looking straight at one another again, after all this time, and what you see in me I really don't know, but I can see whatever horrible machinery was gearing behind them, the brilliant light of your eyes remains.

And we are both looking through that light, looking without blinking like the children's game: I am leading you back out of the clinic room, glissading then tip-tapping down the corridor past 'Pain', 'Epilepsy', 'Multiple Sclerosis', 'Neuro-oncology', across the noisy waiting room, the camera still following, leaving this building after more than twenty-four years in A&E, out into the middle of the day.

I hail a taxi and we get in.

'St Pancras.'

From there to Paris, Marseilles by midnight ...

Chisel ... aqua ... mandarin ... Istanbul ...

Me

Day 1

'Miss Amrita Chakraborty?'

'Miss Elisa Garcia De La Huerta?'

'Mr Ron Nehemiah?'

'Dr Alas-ter Ben-yamen?'

He calls out our names into the reception area, quietly, almost whispered, respectful of the silence.

'Dr Ben-yamen?' He means me.

There is no handshake, no eye contact, no small talk about my journey. I follow him in silence down a pavement lined with dying trees. The jungle is landscaped in such a way that there is not one vista of more than a stone's throw, to restrict seeing, to turn one's back on the future. In one corner of the property is a large stupa – a greying white shrine mounted with CCTV. There are new-build cottages – 'kuttis' – dotted throughout, a larger hall, a refectory. All the buildings are sinking, the moss and lichen already up to the bottom of the windows. His is the last on the left.

We sit outside his kutti on two wooden stools facing one another under the shade of a large banana tree: his clinic room. He wears a dull amber robe, his head is shaved. He is older than me, of indeterminate Asian origin, Burmese perhaps, or it could be Tooting.

The moments go by. No charge passes between us, nothing except mosquitoes fizzing half drunk in the hot, sunless air.

A flight yesterday morning, another flight, then a train, a night bus, a short walk across the border, a taxi, and when the taxi driver got lost a cycle rickshaw. I am dropped at dawn by the perimeter gate. Through barbed wire, the silhouettes of white toy-town temples – from Sri Lanka, Bhutan, Tibet, Thailand, Japan – like outsized wedding cakes in a bake-off for Asian billionaires. A quagmire recently declared a world heritage site, commemorated with future relics, sinking before complete, then an earthquake, so that now these brief architectural dreams are semi-derelict white elephants turning green in the swamp.

There is a board with written instructions: fourteen hours per day of sitting and walking. No lying, no stealing, no sex, no phones, no books, no talking, no writing. Just empty corridors, CCTV, and slow-moving people not making eye contact. Same as any psychiatric ward.

There is a clock that gives the temperature and humidity: 41°C/99%.

In the hall the other meditators are cocooned in individual mosquito nets which drop from the ceiling, breathing in and out like lava in giant chrysalises, waiting for liberation. Heat accelerates this, but hatch too quick and they die.

Fourteen hours? *Fourteen?!* Recreation, love, spirituality – each turned into *work*: this is how we cope …

A current darling of neuroscience research – the cultivation of default-mode networks – indicates that our brains need mindlessness, unemployment, fucking about, eating mental crisps, in order to thrive.

I manage to sit for fifteen minutes before I change position, distracted by an involuntary thought of the beautiful Miss de la Huerta. As soon as I change once, I keep changing: wriggling, itching, sniffing, coughing, writhing ... Movement is suffering in action.

It is dark now. The clock says 38°C/99%. The fan doesn't work. The light doesn't work. I sit in the dark sweating. Nobody knows I am here.

Day 2

Dawn approaches. I didn't sleep. A digital bass gong rang every thirty minutes through the night. There is chanting now, either real or piped.

Later. I tried to call the kids on WhatsApp using the phone I didn't hand in.

'Daddy ... ? Daddy ... ?'

The reception is terrible.

'Are you there sweetheart ... ? Darling ... ?'

Together again under the banana tree. The monk's face is hollow, pointless as stone. I look as deeply as I can, but it gives back little: Alzheimer's probability 18%, exceptional reserve from years of meditation practice; vascular dementia 12%, reduced by vegetarianism, low body fat, no stress.

'Plunge into the object. Note the feel of your breath, rising and falling, rising and falling ...' He speaks in perfect Downton

Abbey. 'Put the heel of your palm on the abdomen.' I do as he says.

We breathe together for a minute.

There is evidence that mindfulness meditation harnesses the brain's capacity for neuro-plastic generativity, producing subtle changes in cerebral blood flow, alpha wave activity, and synaptic connection and disconnection rates following brief regular practice.

There was a desperate crush on mindfulness, a superficial emollient for our late-stage attention-cancer. Never mind the patients, it might save the health professionals. Even the surgeons were at it: meditating with eyes shut in surgery – helped them drill straight. And it was rife in clinical psychology, for the mass of clinical drones slaving in the Siberian salt mines of protocolised CBT delivery: average life expectancy in the mid-forties.

Day 3

We sit opposite one another. I shaved my hair off like him this morning. I thought it would make me feel different. I did feel different. Then five minutes later I felt the same.

'Keep returning to the breath. Storm it ... rush it like you would rush an urgent patient to hospital.'

His words are lightly slurred. Possible differentials include Primary Progressive Aphasia vs Ischaemia of Middle Cerebral Artery – both of which compromise speech production.

Tomorrow, 5,000 miles away, was to have been our engagement party, if it wasn't today. I will be alone forever ...

'Don't be distracted by these thoughts. Just the rising and falling of the breath.'

As a child I spoke when I thought. I thought that had stopped.

'It hasn't,' he says.

★

Due to its inherently internal, private nature, mindfulness meditation practices are naturally difficult to observe or control methodologically.

We, my line manager and I, had agreed that it would be called a 'sabbatical'. There were good, evidence-based reasons for going, new papers every week on the neurobiology of meditation and its therapeutic applications. I would try and write something up: 'Hospitalising the East' was a working title.

I am not supposed to be writing. We cannot tell our stories and be present at the same time.

Day 4

The room is his not mine. I keep looking at him for symptoms where he sees through me.

'The face is just a mask. There is nothing underneath,' he says.

I have a long psychiatric history that I didn't declare on my application form.

No lying.

'Awareness is the true face of the mind,' he says.

I am finding it hard to breathe here.

'What's happening?' I ask him.

'Nothing.'

Silence. Then,

'If the breath is shallow, note it. If it is deep, note it.'

Day 5

The clock reads 41°C/99%. It must be broken. Unless nothing changes.

I am the worst here, worst because nobody else apparently cares, or knows how good they are, or what good might mean, or even that there are others here at all, who they might or might not be worse than.

Different meditation traditions are supposed to activate different neural networks. Hindu-inspired meditations typically demonstrate a left-hemisphere lateralised pattern of neural activation, whereas Buddhist-inspired meditations, which are more akin to mindfulness-based practices, are thought to harness brain activity that is primarily anteriorly focused (Tokley 2018).

A 'sabbatical'? She was sacking me in the only way she could. I had signed the paperwork without reading it.

I abandon the research.

Day 7

'The organism has a definite number of breaths'; 'Speed accelerates death: slow down your breath, stretch out your life.'

This is really how he speaks, with the pronunciation of minor royalty.

'Where are you, Daddy?' Bron is asking.

I am trying to explain, but my words aren't coming out, or they are coming out and not reaching her.

'But where are you?' Her voice is urgent.

I am not there. Fatherhood is crucial in the later punctuation of preadolescent attachment behaviours (Bron's age), especially in girls – crucial to forming resilience, well-being, self-efficacy, helping their own prospects as parents.

Mould has formed on my clothes in the humidity. I search through a basket of cast-offs. I find blue scrubs – perhaps the abandoned uniform of some rogue intensivist. A mask for the pollution, gloves to protect against germs: one day we will all dress like surgeons.

'No Scrubs' (TLC 1999). I choose a white shirt and sarong. A monk again.

★

Occasionally I see a local pruning the jungle-garden; barefoot, threadbare trousers — one of them wears an Arsenal shirt — speaking quietly into large 4-Sim 5G phones.

I hand my phone in. I keep still. Nothing changes. I label the rising and falling of my abdomen as I have been told. I label distractions for what they are — 'Feeling Feeling' 'Thinking Thinking' (I notice I am labelling things with a posher accent than is normal for me, his voice channelling through me). Within a minute, less, I forget my purpose. Nothing stays the same.

Day 8

Spend most of the day in my *kutti*, sweating, crying, incontinent, unable to get out of bed.

Kafka's story ends with the doctor in a state of permanent hellish flight, speeding through the winter landscape on his troika, at the mercy of his demonic horses, bitten by the cold, unable to reach his fur coat which hangs from a hook, just out of reach. He has abandoned a patient, a dying patient, having misdiagnosed him, having summarily dismissed his family, having lain down in the patient's bed in resignation, taken up the space of a dying man. And there, next to him, the patient, by turns non-compliant, awkward, vague, suggestible, a symptom-amplifier, worse still a malingerer, hostile, resentful, hopeless, is dying …

I am both.

Day 9

A terrible burning sensation below my scapula within two minutes of sitting, like someone has sewn a leaking hot water tap there. Pain threshold of a child.

Later, in front of him again:

'What are the details of your suffering? When does it start? Where? How? What do you notice?' Just like my clinic, same questions, same hunt for the killer detail.

'Get close, as close you can bear. Then get closer.'

Close? 5,000 miles away – my daughters. The only thing that really mattered – not to do to them what had been done to me – and I did it, over years, by a sequence of abandonments. Life, a lurching from crisis to crisis, because crisis is the only pulse I know; each episode an apparent accident, a random word unrelated to the last, but really an unfolding sentence, a life sentence, predictable from the beginning.

'Your thoughts are just noise, judgemental noise,' he tells me. 'You are like a dog, a dog barking up the wrong tree, because there is no tree.'

I had a dog once, a beautiful, deranged Anubian Alsatian which I called 'London'. She was my longest relationship. We would spend whole days together picnicking in Hyde Park. Sometimes she would disappear in search of squirrels, and I would spend hours calling for her, standing in the middle of a crowded park, hands megaphoning my mouth, screaming, 'LONDON! … LONDON! … LONDON! …', no dog in sight. Everybody knew what I meant.

No words, no words: hush.

'They get in the way, just the rise and fall of the abdomen.'

Day 10

Breath stops. I wait for it to come back.

'Treat every breath as your last.'

'I don't feel well.'

'Just note it: "Feeling Feeling". Then move on to the next moment,' with that curt, doctorly diction.

Another father who won't hear me. But I really don't feel well.

I imagine 'making up' a symptom, but even in fantasy I am in his room, stuck in his gravitational field: I can't do it.

Day 12

Rising and falling, rising and falling. Over and over and over …

Boredom doesn't stay boring, it turns into something more threatening (Feeling Feeling). Waiting for my father's last breath. Waiting for Cordelia's first breath. For Bronwen's to return after her seizure … Can you die on an in-breath? Can you be born on an out-breath? (Thinking Thinking.)

Day 13

I notice each rising is slightly different, each falling is slightly different – as he predicted.

Back pain starts to dissolve very slowly if I don't move and keep watching it, just as he predicted. And when I keep watching it, it keeps changing, moving towards its opposite. Pain. No pain. I think how this internal voice – although 'voice' is far too coherent for what it really is – has a concept with the label 'pain'. Even though it is out of date, describing a non-painful experience, I continue to use 'pain' and think in its terms … Concepts strangle the truth they try to tell.

'I' am not there, he tells me. Just pain: and that's not there either. 'I' – the ur-concept: really only pain pains; anger angers; madness maddens.

I focus on the abdomen, but I cannot find it, not with my mind, not with the palm of my hand. I am not alarmed. There is light joy. I imagine subtle re-innervations in the somato-sensory map as they take place (Thinking Thinking). I think of dopamine and serotonin flooding the synaptic clefts as it happens (Thinking Thinking).

It's just as he predicted this morning:

'There may be dislocations in your self-experience.' He is gnomic. 'You might even find them pleasurable,' gnomic with killer dryness.

He creates the expectation and this changes how and what I notice of the physical world. Meanwhile it changes the physical world at the same time. He is the placebo, his performance note perfect, with me his suggestible subject.

'Still thinking thinking,' he tells me. 'Your thinking gets in the way of knowing.'

Day 15

People only arrive here, nobody leaves. Like an asylum. The future captains of industry will be endurance meditators: they will understand the mind of everyone having encompassed all experience with their own minds first.

Day 17

Long inhale, long exhale. Like a silent spondee.

I become aware of expectations of the breath, a rhythm which I am always slightly ahead of, by just a fraction of a moment.

I play with it, imagining the shape of the abdomen like an incinerator with a tall chimney stack leading up to my mouth; the rising and falling like an unsafe hospital building being detonated and then momentarily reassembled; I imagine breath rough as sand and it coarsens, or like it's raining cats and dogs and they fight in front of me, as colour, memory, love, whatever comes to mind: language, similes, visualisation; these are the cortical dictators of our sensory experience. I am the conductor, the breath my orchestra.

This is high-grade brain-training. I can feel neurons integrating. If I can keep it up my cerebral reserve will be bottomless.

Day 21

Sleep down to two to three hours per night. I walk five yards, it takes forty-five minutes. The level of detail in every step is thrilling, phenomenally overloading. Ecstasy for a moment, then sadness, that so much is unnoticed, so much of my life has gone by without my noticing it, missed, autopiloted, unselected, lied to.

Can you have a breakdown in a breakdown?

What if everything is breaking down, always breaking down? A self-portrait in a convex mirror that has been smashed to smithereens.

Day 26

We face one another, him and me, mind and brain. He is what he is doing with the rest of his life.

A long time ago, amidst another crisis, the monks had sent me to Dr Lotte. She was in her late sixties even then, but would be youthful on the day she died. Austrian by birth, she had lost her family in the Holocaust, was refugeed to Shanghai where she was well educated in the Jewish ghetto, then refugeed a second time to LA and the Hollywood ghetto. It was 1947. She got a job as a receptionist at MGM where she met Alan just back from the war – he'd arrested Goebbels, held a gun on him, imagined what it would be like to pull the trigger – was writing a film with Thomas Mann ... *Thomas Mann*! And in the evenings he was a jazz pianist and a political radical, at weekend pick-up games a no-nonsense point guard ... This was a time when 'soulmate' still really meant something. Lotte had the keys to the studio lot so they kiss on their first date on the fine white verandah overlooking the grand Southern ranch mansion in *High Society*; they go for moonlit walks by the Seine in *An American in Paris* (the fake river, no deeper than a kiddies' paddling pool); they play hide and seek in *The Big Sleep*'s orchid house ... Or that's how she remembers it. Later, tired of movies ('For the developmentally delayed' according to Alan) they head off to Mexico where they build a hospital for local farmers.

'Why are you telling me this?' I ask.

'You'll see,' Lotte says.

'But why are *you* telling me this?' the monk asks.

'Because I was a monk like you, thought I could do it for the rest of my life – until I didn't.'

In Cuba Alan trains as a medical anthropologist and Lotte starts her training as a clinical psychologist. Over the years they travel round South and North America in a camper, home-schooling their children, treating indigenous populations from mobile clinics, generally empowering and radicalising the communities as they go, depending on their needs, and in their spare time writing string quartets and little two-handed comedies for the kids to stage ...

I'm telling you all this because I was so impressed by the outside of someone's life, their attainments (forgetting that nine-tenths of it is lived on the inside) that I let my vocation hinge on a wish that was not even mine, the next two decades decided by the momentary intersection of two strangers' fantasies. That first session Lotte told me the story of her life. Over the next few sessions she listened to my high-wire blather, looking about as wise as it's possible to look. Then she said: 'Here's what you're gonna do with the rest of your life ...' Simple as that, like the Wizard of Oz. I'd already been told the answer to Murray's question. Then I forgot.

I did exactly what she told me to ...

'What have I done with my life?'

'You treat yourself as solid when no such solidity exists,' is all he says.

'But what's happening?'

'You are.' His eyes spying me from behind a painting.

Day 34

It is my forty-eigth birthday today, my mother Lucy's seventy-second: I'm exactly two-thirds of the way to her and closing.

The fatal neurological calendar that Lewis and I cooked up should extend its scope beyond the diagnostic moment: indicate the age at which the first pathogens or biomarkers of specific diseases appear – say, at forty-eight: precursor cells nestling in the substantia nigra that in twenty years will be as diagnosable as Parkinson's. The future is certain; give us time to work it out (Byrne 1982). Or we could make birthday cards that detail the number of other forty-eight-year-olds on the planet who were born on the exact same day as you who are still alive, and then the number of their dead others, giving their current longevity a percentile rank.

Mindfulness not for stress relief, improved focal attention, thinking straight, but for letting go, each breath experienced as though it were the last: mindfulness for death and dying.

Day 35

Liberation. I want him to be right because I fear I might be immune, unreachable, unchangeable. He wants to be right because he's spent forty years of his life doing this, believing that every one of us is teachable, that we might see whatever light he sees; that's a long time dedicated to getting something wrong. Like the doctor and his patient, our intentions lead us to try and coincide, collaborators on a single story.

I say things about my experience, he tells me if they are authentic or not. In this way he schools me in his language for who I am: 'suffering', 'impermanent', 'self-addicted'. How different if the room were mine: the slow deliberate meditation walk becomes the shuffle of Parkinson's; the tingling sensations the meditator feels in his limbs – progressive supranuclear palsy; the disintegrating body map of the experienced yogi – severe TBI? Lewy Bodies? His disorientation – post-traumatic amnesia? Late-stage Alzheimer's? Auditory hallucinations, thought disorder – neuropsychiatric prodromes … ? Only here, in his room, they are stages of enlightenment.

Day 37

Still writing although it's prohibited, still the same urgent, undirected prolixity, the need for words, words, words. Syntax getting balder, shiny like my shins after four and a half decades of rubbing against trousers. Vocabulary thinning, punctuation on a ventilator ... A colleague made his name detecting thinning of written language as a preclinical sign of Alzheimer's, before even the author is aware.

I hear everyone back at the hospital is working on a book or a script. Nobody is quite *there* any more. Even though the TV crew has packed up and left, the damage has been done; everything has become potential material, shoved into the space between quotation marks, our lives no longer simply our own.

Day 41

You breathe in, you breathe out, you breathe in, you breathe out ... until you don't. *Why should a dog, a horse, a rat, have life, and thou no breath at all? ... Look on her, look, her lips, look there, look there!*

Day 47

41°C/100% humidity: am I breathing underwater now?

The place is sinking into the swamp a little more each day, like being reclaimed by fiction.

Our eyes lasso each other across the room: your open featureless face, ready to register anything that moves, endlessly responsive. My breathing shallows, my heart rate increases. I note that I feel love for you (Loving Loving) and with it the seed of the next loss.

'It is clinging, not love,' you tell me.

Day ?

Birds squawking, ripple-less in the still lake of my attention ...

We've exhausted our brains. It won't be long before people stop caring about the location of love, the pathway for financial success, the area associated with gifted golf swings. There will be a few wizened neuroscientists, undistinguishable from the rest of the homeless, keeping warm around a campfire built in a disused MRI coil, telling each other stories about 10 tesla tractography and the Binding Problem. (Thinking Thinking!)

We are sinking, right now.

'Madness', *the* quintessential symptom of selfhood: our compulsion to constantly confirm ourselves, bring to life that ever-vague idea, like pulling back the curtains on the Great and Powerful Oz to find a doddering old man, who confesses that he can't give you a heart or a brain, just a clock or a degree certificate; only for another curtain to be pulled back behind him, to find another even older man, a still less wizardly wizard, who can't even give you a clock or a certificate, or even his name, only blank paper for you to write down what you want from him; and on and on, curtain after curtain, old man after old man – maybe they come to look more and more like your father – until there is a naked nonagenarian with dense Alzheimer's who no longer blinks; until there is nothing behind the curtain.

Let me not be nothing.

'Madness'. There is no stopping our searching: subtly, in environmentally wrought imbalances of neurotransmitters; in prenatal-perinatal-postnatal anomalies; in delinquent mitochondria; deeper still, in predisposing gene sequences, disinhibited protons, subtler and subtler, until almost nothing, until nothing. Or more grossly, extrapolating from adverse developmental contexts, to unstable intimate relationships, frayed social networks, community prejudices, socio-economic hardships, grosser and grosser, racial strife, civil war, international trauma, continental drift, alien insertions, universal sadness, until everything.

Really, there is no constraining the elaboration of this search in either direction; every gap in our experience must be filled and explained, until it exceeds the point of maximum complexity. We are exactly as lost as each other: even the most particular is happening to everyone and the most general we have made our individual signatures. Meanwhile that which isn't us is taking over, feasting on our attention without our knowing, automating our most intimate feelings, helping our brains to find new ways to dement, illegible to the most advanced and penetrating instruments.

'Sanity' is nothing more than the shrinking back from this complexity, a whale-sized Russian doll that will only house what looks like it; emptiness, nothing, like white light, meaning all the madnesses combined. And this giant Sanity Doll carries us along, making our decisions for us, so we have no influence on where we are going, and if we resist we are ejected violently like unwanted progeny, made to go it alone, left to scream our names over and over again in a language nobody else seems to speak ... Until we exhaust ourselves, and return to the crowded waiting room to beg for readmission ...

Day 60-something

Weightless, an astronaut inside my own body, my spacesuit filled with light. (Thinking Thinking.)

'There is no "body". There is no "inside". There is no "my".'

Day ?

They have followed me here; Michael, Jonah, Tracy, Jane, my patients ... I disappear completely but they find me anyway.

Day ?

Supposed to leave last week, but when I tried, my bag refused to be packed.

Sit effortlessly for hours. Day turns to night. I forget to move.

Day ?

There is nothing but arising and passing away. Breath is like ... nothing else. Words never surpass the bliss of breathing.

Place hand through head: no brain, no mind, no hand.

Day ?

Endings dominate. Nothing that forms isn't already falling to pieces. Falling Falling.

Day ?

And.

Day ?

' ... '

' ... '

'*Do you know me?*' says Cordelia.

'*I fear I am not in my perfect mind ... I think this lady to be my child Cordelia.*'

For a moment she and Bronwen are sitting next to me, approximating ancient Egyptians.

Epilogue

Treatment history: AKB.

Child episodes:

One referral to child psychologist, only parents attended. Otherwise no recorded engagements with child and adolescent mental health services.

Adult episodes:

1. Six months of cognitive behavioural therapy under psychiatrist - outpatient at a day unit following apples apples. 1989. Proposed diagnosis of apples apples disorder.

2. Three months of psychodynamic psychotherapy, following probable co-morbid apples and apples. 1990.
3. Ten weeks of counselling following apple bingeing including an apple seizure and possible apple psychosis. 1991.
4. Three weeks of counselling following attack of apples and apple ideation. 1996.
5. Eighteen months of weekly humanist/integrative psychotherapy for acute apple, plus chronic apple with co-morbid apple problems. 2000–2.
6. Four sessions with clinical psychologist Dr Lotte M. in California, described as 'vocational/spiritual discernment'. 2004.
7. Six months of weekly psychodynamic psychotherapy to cope with apple issues. Diagnosis of possible apple personality. 2005.
8. Six sessions of mindfulness-based therapy for apple management following the start of a new relationship and pregnancy. 2006.
9. Two years of Kleinian-oriented psychoanalysis, four/five times per week, in the context of becoming a father. Closely followed by breakdown of new family. Provisional diagnosis: acute apple with chronic apple existing on Axis II, at the level of his personality. 2006–8.
10. Nine months of couples' therapy with the children's mother under a clinical psychologist. 2007–8.
11. Jungian psychotherapy on an intermittent basis, 2012–14. Including necessary switching of analysts because of therapist's apples diagnosis.

12. Two sessions with Jungian psychiatrist. Patient requested immediate ending and that fiancée be seen in his stead. Patient conceded two decades of intermittent attendance at four different types of Twelve Step fellowship.

13. Pharmacology. Twice been placed on apple medication (Apple, Apple), both times self-requested. On one occasion anti-apples were added to mollify anger. On both occasions medication was aborted within two weeks, before physiological drug effects were possible.

Drinking in some London pub, watching OJ Simpson live on CNN in his white Bronco running from the cops down Sunset Boulevard where the bars had people glued to CNN *live* – looking as drunk as I was – then running in delirious excitement out onto the sidewalk – live *on* CNN – to watch OJ pass by – *live on* CNN (I nearly stuck my head out of the door of my pub, just to check), and then look back to the TV again, as though they couldn't quite decide which version was the most compelling. Well, that moment confirmed how screwed things were: unfolding forever shot through with its own commentary. From that point on our lives would be *based* on actual events and we would have to perform our own stunts.

It is still a dismal-looking afternoon in London in November because it always is; soulmates waiting for one another in acclaimed coffee shops, unclaimed children hanging around school gates, a white-faced man Everesting on Primrose Hill. He stops halfway up, heart marooned in zone 5, closer to death than he knows. It starts to spit with rain, meaning there will be a dozen extra head injuries before the day is done.

Confessions, if they're worth anything at all, are painful, partial, protracted ... In many ways I'm still clearing my throat, between stories; the end of one chapter of madness, the beginning of another. Change is long and torturous, 700 pages long and called *Crime and Punishment*, not *Crime and Redemption*. Unless, that is, it's sudden, turning on a single word, apparently unprecedented, our hero only seeing something like light – which may just be darkness pausing – on the final page. And that is not withstanding the problem of wanting change when really this is also wanting not to be yourself, if indeed there is anything to change into, if there is anything to be done at all.

As for you, dear patient, never forget we care. Remember it – as long as the faculty for memory remains intact – when you are next looking across the table at your doctor wearing his nice suit, an eccentric bow tie or scrubs, looking straight back at you, meeting your question with gentle, light, admiring, now fierce, dark, delinquent eyes, and you find yourself wondering what, if anything, he is thinking.

Select Bibliography

Bollas, Christopher, *Being a Character: Psychoanalysis and Self Experience* (Routledge, 1993)

Bollas, Christopher, *Cracking Up* (Routledge, 1995)

Crittenden, Patricia M., *Raising Parents: Attachment, Parenting and Child Safety* (Routledge, 2008)

Eigen, Michael, *The Sensitive Self* (Wesleyan, 2004)

Eigen, Michael, *Faith* (Routledge, 2014)

Tokley, Melanie, 'How the neurobiology of mindfulness and other awareness-based meditative practices from the Eastern wisdom traditions might be clinically relevant. Findings from brain imaging (MRI, fMRI, PET) and electroencephalography' (2018)